FARM MOTORS

FARM MOTORS

PRACTICAL HINTS FOR HANDY-MEN

by

J. Brownlee Davidson

and

Leon Wilson Chase

FOREWORD BY NATHANIEL TRIPP

THE LYONS PRESS

Originally published in 1908 by Orange Judd Company, New York

First Lyons Press edition, 2000

Printed in Canada

10 9 8 7 6 5 4 3 2 1

Library of Congress Cataloging-in-Publication Data available on file

CONTENTS

FOREWORD

It began with the wind, and the notion that perhaps the wind could do more work than simply separate grain from chaff. As the farmer's labor became less physical and more cerebral, systems of rotating sails were devised and constructed into the earliest of farm motors. There was also the fancy that perhaps the restlessness of farm animals could be harnessed, too, harnessed not to pull but to turn. Although the exact times and locations of these early inventions are lost, it is known that as the Roman Empire sank deeper into decay, these earliest of farm motors were beginning to revolve. In their slow, creaking way, they were starting a transformation of the world more profound than the Roman Legions themselves could have imagined.

Farm Motors begins with these early contraptions. Although they may not fit our modern notion of a motor, they still fit the definition by converting one kind of energy into another to produce motion. They can still be found around the world today, pumping water and processing crops. The windmills stand out particularly for their grace and variety of design, from the classic Dutch to the mass-produced American models that dot the Great Plains. Far less graceful and efficient are the ones cobbled out of boards and boxes, but a pioneer in the open lands of the West could put one together for next to nothing.

Windmills, however, keep their own hours, performing their work at the weather's whim. They are best suited for pumping water; the work they do is stored for later use. The clattering treadmill, using a horse or two, is better for processing crops. Part of the science lay in knowing the horses themselves, recognizing a good one, and knowing how long it could endure plodding along a continuous belt on an inclined plane or, less efficiently, circling at the end of a spoke. Unlike the wind, a horse would work at the farmer's command, but then they kept on consuming the latent energy of grain and hay, day and night, long after the work was done.

In 1900 there were 18 million horses and mules on American farms, but agriculture was poised on the verge of another revolution. The notion of horsepower would change forever when the laws of thermodynamics were applied to the laws of motion. At last we had motors that worked only when we wanted. Whistling louder than the wind itself, steam power was blowing across the land through the nineteenth century. It was so powerful and so complex that it was late in coming to the farm; there was not yet enough work for it to do. Besides, running a steam engine required a degree of esoteric knowledge on the part of the engineer comparable to that required to run a nuclear power plant today, and with similarly tragic consequences if mistakes were made. Nonetheless, these monstrous motors were not only powering bigger and bigger machines, but they were even beginning to lurch and wheeze across the fields on their own.

It is my dream to encounter one of these, awaiting my whim, in person. Everything I need to know to get it underway is here, from kindling the fire in the morning to dropping the ashes at night. With sliding valves and eccentrics, governors and injectors, this is a technology that had reached a level of enormous complexity while still depending upon the art of judging the soundness of a boiler by the ring of hammer blows—or the timing of the steam cutoff by the bark of the exhaust. Such engines had a great appetite for fuel—burning corn cobs, straw, wood, or coal—and an even greater thirst. Especially fascinating is the section on "Handling a Traction Engine." Nowhere else will one find such mechanical melodrama couched in the succinct prose of a manual. Going up a hill required considerable thought and preparation beforehand.

Descending the other side was simply terrifying. The traction engine was much better suited to running on rails and would spend a heroic century there, while only making a fleeting appearance on the farm.

The authors themselves recognized the limitations of steam traction when compared to the potential of the internal combustion engine, which was still rapidly evolving at the time. "In proof of the fact that it may be made to furnish its own tractive power it is only necessary to refer to the automobile, which is made to work under great variance of speed," they said prophetically. It was with the internal combustion engine, with its greater thermal efficiency, its simplicity and versatility, that the farm began to move to the syncopated rhythm of the times. As always, this was an awkward beginning; the only effective muffler for the first stationary gas engines, for example, was a small room built of brick.

Electric motors were beginning to find their way to the farm at this time as well, adding their hum to the beat and the throb of modern agriculture. Their portability adapted them to many tasks, but the wires connecting them pointed to another profound change taking place on the farm. While the farmer was feeding more and more people, he was also less independent. He had to feed his motors fuel and lubricants and electricity produced at a great distance, and was dependent upon the latest information as well. He was becoming an integral part of our industrial society, woven into the woof and warp of the times.

There was an endless fascination with the clicks and whirrs and pulsations of farm motors on the part of a boy born to a Minnesota farm in 1902. The young Charles Lindbergh learned of their inner workings by taking them apart and marveling at the complexity of their engineering. He plowed with a two-cylinder, three-wheeled monstrosity built by La Crosse, and became the local agent for the electrically powered Empire Milking Machine while still in his teens. Sales were slow; local farmers feared the machines would damage their cows, but the young man was undaunted. He bought an Excelsior motorcycle. He took a flying lesson. Just within the first two decades of his life, he was witness to the enormous changes in farming wrought by motors, yet it had all begun with the wind. And when he turned and left the farm, as so many young men did, he was borne upon the shoulders of the wind itself.

—Nathaniel Tripp
St. Johnsbury, Vermont, 2000

FARM MOTORS

INTRODUCTION

Motors.—The application of power to the work of the farm largely relieves the farmer from mere physical exertion, but demands of him more skill and mental activity. At the present time practically all work may be performed by machines operated by power other than man power. This change has been important in that it has increased the efficiency and capacity of one man's work. Farm Machinery has been a discussion of the machines requiring power to operate them, while Farm Motors will be a discussion of the machines furnishing the power. The number of machines requiring power to operate them is increasing very rapidly. They require the farmer to understand the operation and care of the various forms of motors used for agricultural purposes.

Energy may be defined as the power of producing change of any kind. It exists in two general forms:

1. Potential or stored energy, an example of which is the energy contained in unburned coal.
2. Kinetic or energy of motion, an example of which is the energy of a falling body.

Sources of energy.—Following are some of the sources of energy available for the production of power.

Potential:

1. Fuel.
2. Food.
3. Head of water.
4. Chemical forces.

Kinetic or actual:

1. Air in motion, or the wind.
2. The waterfall.
3. Tides.

The energy found in the forms just mentioned must be converted into a form in which it may be applied to machines for doing work. This change of the energy from one form to another is spoken of as the transformation of energy.

The law of transformation of energy holds that when a definite amount of energy disappears from one form a definite amount appears in the new form, or there is a quantivalence.

Prime movers are those machines which receive energy directly from natural sources and transmit it to other machines which are fitted for doing the various kinds of useful work.

Forms of motors:

1. The animal body.
2. Heat engines—
   - Air,
   - Gas or vapor,
   - Steam,
   - Solar.
3. Water wheels.
4. Tidal machines.
5. Windmills.
6. Electrical motors.

Of the above all are prime movers except the last named, the electrical motors. The energy for the animal body is derived from the food eaten. This undergoes a chemical change during the process of digestion and assimilation, and is transformed into mechanical energy by a process not fully understood. Heat engines make use of the heat liberated by the chemical union of the combustible constituents of fuel and oxygen. Water wheels, tidal machines, and windmills utilize the kinetic energy of masses of moving water or air. Electrical motors depend either upon chemical action or a dynamo to furnish the energy, it being necessary to drive the latter with some form of prime mover.

Only such motors as are well adapted to agricultural purposes will be considered in this treatise.

CHAPTER I

ANIMAL MOTORS

The animal as a motor.—Although the animal differs from other forms of motors, being an animated thing, it is possible, however, to consider it as a machine in which energy in the form of food is transformed into mechanical energy, which may be applied to the operation of various machines. The animal as a motor is exceptionally interesting to those who have made a study of the transformation of heat energy into mechanical energy, for this is really what takes place. Combustible matter in the form of grain and other foods is consumed with the resultant production of carbon dioxide or other products of combustion in various degrees of oxidation, and, as stated before, mechanical energy is made available by a process not clearly understood.

Viewed from the standpoint of a machine, the animal is a wonderful mechanism. Not only is it self-feeding, self-controlling, self-maintaining and self-reproducing, but at the same time is a very efficient motor. While the horse is like heat engines in requiring carbonaceous fuel, oxygen, and water for use in developing energy, it is necessary that combustion take place in the animal body at a much lower temperature than is possible in the heat engine, and a much smaller proportion of the fuel value is lost in the form of heat while the work is being done. The animal is the only prime mover in which combustion takes place at the ordinary temperature of 98° F. For this reason the animal is one of the most efficient of prime movers. That is, a large per cent of the energy represented by the food eaten is converted into work, a larger per cent than is possible to realize in most motors. Professor Atwater in his recent experiments found the average thermodynamic efficiency of man to be 19.6 per cent. Experiments conducted by the scientist Hirn have shown the thermodynamic efficiency of the horse to be about 0.2. The best steam engines give an efficiency equal to this, but the average is much below. Internal-combustion engines will give a thermal efficiency from 20 to 30 per cent.

Muscular development.—It is possible to consider the animal as a motor, but the animal is made up of a great number of systems of levers and joints, each supplied with a system of muscles which are in reality the motors. Muscles exert a force in only one way, and that by shortening, giving a pull. For this reason muscles are arranged in pairs, as illustrated by the biceps and triceps, which move the forearm. It is not clearly understood just how muscles are able to exert forces as they do when stimulated by nerve action. The theory has been advanced that the shortening of the muscles is due to a change of the form of the muscular cell from an elongated form to one nearly round, produced by pressure obtained in some way within the cell walls. There is no doubt but there is a transformation of heat energy into mechanical energy. While at work and producing motion there is but little change in the temperature of the muscles, but when the muscles are held in rigid contraction, there is a rise in temperature. Another author* has likened this to a steam plant, which while at work converts a large portion of the heat generated in the fire box into mechanical energy, but as soon as the engine is

*F. H. King, in "Physics of Agriculture."

stopped and the flow of steam from the boiler stopped the temperature rises rapidly.

Strength of muscles.—All muscles act through very short distances and upon the short end of the levers composing the animal frame. Acting in this way speed and distance are gained with a reduction in the magnitude of the force. A striking example of the strength of a muscle is that of the biceps. This muscle acts upon the forearm, while at a right angle with the upper arm, as a lever of the second class, with a leverage of 1 to 6. That is, the distance from the point of attachment of the muscle to the elbow is but one-sixth of the distance from the hand to the elbow. A man is able to hold within the hand, with the forearm horizontal, as explained, a weight of 50 pounds, necessitating an exertion of a force of 300 pounds by the muscle. Attention may also be called to the enormous strength of muscles of a horse as they act over the hock joint while the horse is exerting his maximum effort, in which case the pull of the muscles may amount to several thousand pounds.

It is because muscles are able to act only through very short distances that it is necessary for them to act upon the short end of the levers in order to secure the proper speed or sufficiently rapid movement.

Animals other than horse and mule used for power.—Dogs and sheep are used to a very limited extent in the production of power by means of a tread power similar to the one shown in Fig. 200 for horses. These may be used to furnish power for a churn or some other machine requiring little power. The use of cattle for power and draft has been practically discontinued in America. An ox at work will travel only about two-thirds as fast as a horse.

Capacity.—A man working a crank or winch can

develop power at his maximum rate. It is also possible to develop power at very nearly the maximum rate while pumping. A large man working at a winch can exert 0.50 horse power for two minutes and one-eighth horse power by the hour. It is stated that an ox will develop only about two-thirds as much power as a horse, owing to the fact that he moves at a much slower speed.

The horse is the only animal used extensively at present as a draft animal or for the production of power. As reported in the Twelfth Census, the number of horses and mules on the farms in the United States was 15,517,052 and 2,759,499, respectively, making a total of 18,276,551 animals. If it be assumed that each animal develop two-thirds horse power, the combined horse power while at work would be 12,184,366, an excess of 184,285 horse power over that used for all manufacturing purposes during the same year, 1900.

From a consideration of the skeleton and muscular development, it is perceived that the horse is an animal specially well adapted to dragging or overcoming horizontal resistances rather than for carrying loads. With man it is different. Although greatly inferior in weight, man is able to bear a burden almost as great as that of a horse, while at dragging he is able to exert only a small horizontal effort, even when the body is inclined well forward. The skeleton of man is composed of parts superimposed, forming a column well arranged to bear a burden. The horse is able to draw upon a cart a load many times his own weight, while he is unable to carry upon his back a load greater than one-third his weight.

It is to man's interest that his best friend in the brute world should be strong, live a long life, and waste none of his vital forces. Much attention has been given to the development of breed in horses. The result is a great improvement in strength, speed, and beauty. But while attention has been turned to developing horses capable of doing better work, few have tried to improve the conditions under which they labor.

That the methods are often unscientific can be pointed out. In England, T. H. Brigg, who has made a study of the horse as a motor, and to whom we must give credit for the preceding thought, states that the horse often labors under conditions where 50 per cent of his energy is lost. It is a very strange thing that men have not studied this thing more, in order that people might have a better understanding of the conditions under which a horse is required to labor.

The amount of resistance which a horse can overcome depends on the following conditions: First, his own weight; second, his grip; third, his height and length; fourth, direction of trace; and fifth, muscular development. These will be taken up in the above order.

Weight.—The heavier the horse, the more adhesion he has to the ground. The tendency is to lift the forefeet of the horse from the ground when he is pulling, and thus a heavier horse is able to use his weight to good advantage. It is to be noted that often a horse is able to pull a greater load for a short time when he has upon his back one or even two men. Experienced teamsters have been known to make use of this method in getting out of tight places with their loads.

Grip.—That the weight adds to the horse's grip is self-evident, but cohesion is not the same thing as grip. Grip is the hold the horse is able to get upon the road surface. It is plain that a horse cannot pull as much while standing on ice as on solid ground unless his grip is increased by sharp calks upon his shoes. A difference is to be noticed in roads in the amount of grip which a horse may get upon the surface while pulling a heavy load. Under ordinary circumstances the improved stone road will not provide the horse with as good a grip as a common earth road.

Height and length.—A low, rather long-bodied horse has much the advantage over a tall, short horse for heavy draft work. He has his weight in a position

FIG. 197—OBTAINING THE WORK OF A HORSE WITH A RECORDING DYNAMOMETER

where he can use it to better advantage. It is an advantage to have the horse's weight well to the front, since there is a tendency, as mentioned before, to balance his weight over his rear foot as a fulcrum. Horses heavy in the foreshoulder have an advantage in pulling over those that are light, as weight in the foreshoulder adds greatly to the ability of the horse to pull. To prove that this is true, it is only necessary to refer to the fact that horses when pulling extend their heads well to the front.

Direction of trace.—A heavy load may be lifted by a common windlass if the pull be vertical, but if the pull be transferred over a pulley and carried off in a horizontal direction the machine must be fastened or it will move. It must be staked and weighted to prevent slipping. This same principle enters into the discussion of the draft of a horse. As long as the trace is horizontal, the horse has to depend upon his grip and his weight only to furnish enough resistance to enable him to pull the load. But if the trace be lower than horizontal the tendency is then to draw the horse on to the ground and thus give him greater adhesion. If the horse has sufficient adhesion to pull a load without lowering the trace it is to his advantage because the draft is often less in this case than any other.

Line of least draft.—When the road bed is level and hard, the line of least draft to a loaded carriage is nearly horizontal because the axle friction is but a small part of the weight.

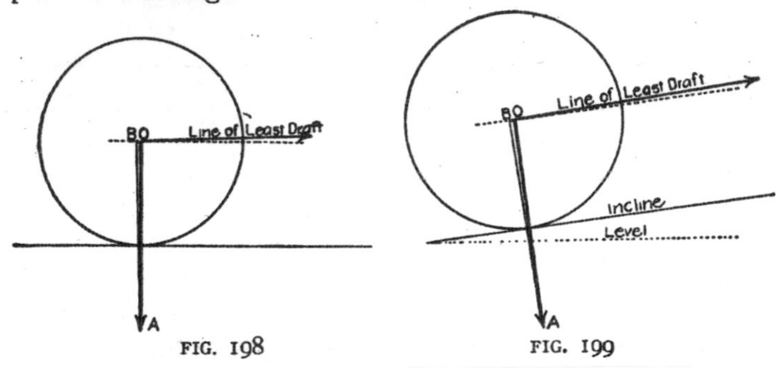

FIG. 198 FIG. 199

Thus in Fig. 198, if AO represent by direction and magnitude the weight upon the axle, and OB in like manner the resistance of friction, the direction of the least force required to produce motion will be perpendicular to AB, a line joining the two forces. The angle that the line of least draft makes with the horizontal is named in mechanics, the angle of repose. If the resistance of friction be that of sliding friction and not that of axle friction, the angle of repose will be much greater.

If the road surface be inclined, it will be found that the line of least draft is nearly parallel to the road surface. If the trace is inclined upward from the line of least draft there is a tendency to lift the load; if the line of draft is inclined downward there is a tendency to press the load on the surface. Furthermore, it is found that roads are not perfectly level and there are obstructions over which the wheels of vehicles must pass, or, in other words, the load at times must pass up a much greater incline than a general slope indicates, and hence this calls for a greater angle of trace than will be needed for level or smooth road. Teamsters find in teaming over roads in one locality that they need a different angle of trace than they find best in another, because the grades of the roads are different.

Width of hock.—As mentioned before (405) practically all of the pull a draft horse exerts is thrown upon his hind legs and for this reason the form and strength of this part must be considered in the selection of a horse for draft purposes. If the hock is wide or, in other words, if the projection of the heel bone beyond the joint is large, the muscles will be able to straighten the limb under a greater pull than if the projection is small; thus the ability of the horse to overcome resistance will be increased. Thus there are many things to be considered in the selection of a draft horse. The general make-up of a horse built for speed is notably different from one built for draft purposes.

The horse at work.—When a horse is required to exert the maximum effort, it is necessary to add to his adhesion or grip so that he may be able to exert his strength to a limit without any slipping or without a tendency to slip. But if the horse is loaded all the time, either by a load upon his back or a low hitch, he is at times doing more work than necessary. In fact, a certain amount of effort is required for the horse to stand or to walk even if he does no work at all. This has led men to think that if the hitch could be so arranged as to relieve the horse entirely of neck weight at times or even raise his trace the horse would be able to accomplish more in a day of a given length. In fact, it might be even an advantage to carry part of the weight of the horse. Although not a parallel case, it is sometimes pointed out that a man can go farther in a day when mounted on a bicycle than when walking. Walking in itself, both for man and beast, is labor, and in fact walking is like riding a wheel polygonal in form, and each time the wheel is rolled over a corner, the entire load must be lifted only to drop again as the corner is passed. Whether or not there are any possibilities in the development of a device along this line to conserve the energy of the horse we do not know; however, the argument seems very good. Mr. Brigg, of England, has devised an appliance for applying to vehicles with thills which will in a measure accomplish the result referred to; that is, the horse on beginning to pull will be gradually loaded down, thus permitting him to overcome a greater resistance.

Capacity of the horse.—The amount of work a certain horse is able to do in a day is practically a constant. Large horses are able to do more work than smaller ones, but a given horse can do only about so much work in a day even if he is given a long or a short time in which to do it. Not only is the ability to do work dependent upon the size, but also upon the natural strength, breed, health, food, environment, climate, adaptation of the load, and training of the horse. A horse with maximum load does minimum work, and when traveling at maximum speed he can carry no load. At an intermediate load and speed the horse is able to do the maximum amount of work.

Best conditions for work.—The average horse will walk from 2 to 2¾ miles an hour, and at the same time overcome resistance equal to about one-tenth or more of his weight. Work may be performed at this rate for ten hours a day. Assuming the above to be true, a 1,500-pound horse will perform work at the rate of one horse power.

As 1,500 pounds is much above the average weight of a farm horse the average horse whose weight is not far from 1,100 will do continuous work at the rate of about 2/3 to 4/5 horse power.

Maximum power of the horse.—Entirely different from other motors, the horse, for a short time at least, is able to perform work at a very much increased rate. A horse when called upon may overcome resistance equal to one-half his weight, or even more. The horse power developed will be as follows, assuming that he walk at the rate of 2½ miles an hour (see Art. 20):

$$\text{H. P.} \frac{1{,}500 \times \frac{1}{2} \times 2\frac{1}{2} \times 5{,}280}{33{,}000} = 5$$

A horse will be able to do work at this rate for short intervals only. The fact that a horse can carry such a heavy overload makes him a very convenient motor for farm purposes.

The maximum effort or power of traction of a horse is much greater than one-half his weight. A horse weighing 1,550 pounds has been known to overcome, when pulling with a horizontal trace, a resistance of 1,350 pounds. With the point of hitch lowered until the trace made an angle of 27° with the horizontal, the same horse was able to give a draft of 1,750 pounds. It is believed, however, that this horse is an exception.

Effect of increase of speed.—As stated before, a horse at maximum speed cannot carry any load, and as the speed is increased from the normal draft speed, the load must be decreased. It is stated that the amount of work a horse is capable of doing in a day is constant within certain limits, varying from one to four miles an hour. Assuming this, the following equation holds true:

$$2\tfrac{1}{2} \times \text{traction at } 2\tfrac{1}{2} \text{ miles} = \text{miles per hour} \times \text{traction.}$$

Effect of the length of working day.—Within certain limits the traction a horse is able to exert varies inversely with the number of hours. When the speed remains constant the traction may be determined approximately by the following equation, provided the length of day is kept between five and ten hours.

$$10 \text{ hours} \times 1/10 \text{ weight of horse} = \text{number of hours} \times \text{traction.}$$

Division of work.—It may not be absolutely true that the ability of a horse to do work depends largely upon his weight, nevertheless it is not far from correct. It is not advisable to work horses together when differing much in size, but it is often necessary to do so. When this is done the small horse should be given the advantage. In determining the amount of the entire load each horse should pull when hitched to an evener it may be considered a lever of the second class; the clevis pin of one horse acting as the fulcrum. From the law of mechanics (see Art. 24):

$$\text{Power} \times \text{power arm} = \text{weight} \times \text{weight arm.}$$

Example: Suppose two horses weighing 1,500 and 1,200 pounds respectively are to work together on an evener or doubletree 40 inches long. If each is to do a share of the work proportionately to his weight, it will be possible to substitute their combined weight for the total draft and the weight of the larger horse for his share of the draft in the general equation and consider the smaller horse hitched at the fulcrum:

$$2{,}700 \times \text{long arm of evener} = 1{,}500 \times 40,$$

$$\text{long arm of evener} = \frac{60{,}000}{2{,}700} = 22\ 2/9 \text{ inches},$$

$$\text{short arm of evener} = 40 - 22\ 2/9 = 17\ 7/9.$$

That is, to divide the draft proportionately to the weights of the horses, the center hole must be placed 2 2/9 inches from the center toward the end upon which the heavy horse is to pull.

FIG. 200—TREAD POWER FOR THREE HORSES

The tread power.—The tread power consists in an endless inclined plane or apron carried over rollers.

and around a cylinder at each end of a platform. Power is derived from a pulley placed upon a shaft passing through one of the cylinders. Fig. 200 illustrates a tread power for three horses with the horses at work. Some aprons are made in such a way that each slat has a level face. This tread is thought to enable the horse to do his work with less fatigue because his feet are more nearly in their normal attitude.

Owing to the large number of bearings, the matter of lubrication is an important feature in the operation of a tread power. Lubrication should be as nearly perfect as possible in order that little work will be lost in friction and the efficiency of the machine may be increased. The bearings should not only have due provision for oiling, but they must be so constructed that they will exclude all dirt and grit.

The work of a horse in a tread power.—A horse at work in a tread power lifts his weight up an incline against the force of gravity. The amount of work accomplished depends upon the steepness of the incline and the rate the horse travels. If the incline has a rise of 2 feet in 8, the horse must lift one-fourth of his weight, which is transmitted to the apron and travels at the same rate the horse walks. Working a 1,000-pound horse in a tread power with a slope of 1 to 4 is equal to a pull of 250 pounds by the horse. This is much greater than is ordinarily required of a horse, but it is not uncommon to set the tread power with this slope. If a horse weighs 1,600 pounds and walks at the rate of two miles an hour, work will be done at the rate of 2.13 H.P. At the same speed a 1,000-pound horse will do 1.33 H.P. of work.

It is often true that a horse will be able to develop much more power when worked in a tread power than when worked in a sweep power, but he will be overworked.

Often horses are overworked in tread powers without the owner intending to do so, or even knowing it.

FIG. 201

Sweep powers.—In the sweep power the horses travel in a circle, and the power is transmitted from the master wheel through suitable gearing to the tumbling rod, which transmits the power to the machinery. Sweep powers vary in size from those for one horse to those for 14 horses. Attention is often called to the fact that a considerable part of the draft is lost because the line of draft cannot be at right angles to a radius of the circle in which the horse walks. For this reason a considerable portion of the draft is lost in producing pressure toward the center of the power, often adding to the friction. The larger the circle in which the horse travels, the more nearly the line of draft will be at right angles to a radius to the center of the circle.

CHAPTER II

WINDMILLS

If the horse is excepted, the windmill was the first kind of a motor used to relieve the farmer of physical exertion and increase his capacity to do work. With the exception of the horse, the windmill is still the most extensively used. To prove that the windmill is an important farm motor, it is only necessary to cite the fact that many thousand are manufactured and sold each year.

Early history—Prof. John Beckmann, in his "History of Inventions and Discoveries," has given everything of special interest pertaining to the early history of the windmill. As it is conceded by all that his work is exhaustive, the following notes of interest have been taken from it. Prof. Beckmann believes that the Romans had no windmills, although Pomponius Sabinus affirms so. He also considers as false the account given by an old Bohemian annalist, who says that before 718 there were windmills nowhere but in Bohemia, and that water mills were then introduced for the first time. Windmills were known in Europe before or about the first crusade. Mabillon mentions a diploma of 1105 in which a convent in France is allowed to erect water wheels and windmills. In the twelfth century windmills became more common.

Development of the present-day windmill.—It was about the twelfth century that the Hollanders put into use the noted Dutch mill. These people used their mills for pumping water from the land behind the dikes into the sea. Their mills were constructed by having four sweeps extending from a common axle, and to these sweeps were attached cross pieces on which was fastened canvas. The first mills were fastened to the tower, so that when the direction of the wind changed the owner would have to go out and swing the entire tower around; later they fastened them so that only the top of the tower turned, and in some of the better mills they were so arranged that a smaller mill was used to swing the wheel to the wind. The turning of the tower was no small matter when one learns that some of these mills were 140 feet in diameter.

John Burnham is said to be the inventor of the American windmill. L. H. Wheeler, an Indian missionary, patented the Eclipse in 1867. The first steel mill was the Aermotor, invented by T. O. Perry in 1883.

The windmills still most common in Europe are of the Dutch type, with their four long arms and canvas sails. These sails usually present a warped surface to the wind. The degree of the angle of the sails with the plane of rotation, called the angle of weather, is about 7° at the outer end and about 18° at the inner. The length of the sails is usually about 5/6 the length of the arms, the width of the outer end 1/3 the length, and the width of the inner end 1/5 the length. It is seen that the total projected area of sails is very small compared to the wind area or zone carrying the sails. Quite often these wheels are 120 feet in diameter and occasionally 140 feet. In comparing these mills with the close, compact types of American makes a very great contrast is to be drawn.

Among the men who have done the most experimenting in windmill lines are Smeaton, Coulomb, Perry, Griffith, King, and Murphy. The names are given in order of date of experimenting. The more prominent among these are Smeaton, Perry, and Murphy. Probably Perry did more for the windmill than any of the others. Prof. E. H. Barbour is noted for his designs and work with home-made windmills.

Home-made windmills.—Professor Barbour made an extensive study of home-made windmills and has had a very interesting bulletin published on the subject. He has classified them as follows:

1. Jumbos (Fig. 202). This type consists of a large fan-wheel placed in a box so the wind acts on the upper fans only.
2. Merry-go-rounds. Merry-go-round mills are those in which the fans in turning toward the wind are turned edgewise.
3. Battle-ax mills (Fig. 203). These are mills made with fans of such a shape as to suggest a battle-ax.
4. Holland mills. Somewhat resembling the old Dutch mill.
5. Mock turbines (Fig. 204). Resembling the shop-made mill.
6. Reconstructed turbines (Fig. 205). Shop-made mills rebuilt.

These mills, although of low power, are used extensively in the West Central States. Most of them are fixed in their position and consequently have full power only when the wind is in the direction for which they are set. In those States in which these mills are used the wind

FIG. 202—HOME-MADE JUMBO

has the prevailing directions of south and northwest, and for that reason the mills are generally set a trifle to the west of north.

To the casual observer the Jumbo mill (Fig. 202) seems a very feasible means of obtaining power, but when one considers the massiveness of the whole affair and that only one-half of the sails is exposed to the wind at one time, also that full power is developed from the wind only when the latter is in the proper direction, it will immediately be seen that only in cases of dire necessity should one waste much time with them.

FIG. 203.—BATTLE-AX WINDMILL

The cost of this type of mill is very slight. It is stated by Professor Barbour that a gardener near Bethany, Nebraska, constructed one which cost only $8 for new material, and with this he irrigates six acres of vegetables. If the water-storage capacity for such mills is enough, they will often furnish sufficient water for 50 head of stock. One farmer has built a gang of Jumbo mills into the cone of a double corn crib and connected them to a small sheller.

The Merry-go-round is not nearly as popular as the Jumbo, in that it is very much harder to build and the only advantage it has over the latter is that a vane may be attached in such a manner that the wind wheel is kept in the wind.

In some parts of Kansas and in several localities of Nebraska the Battle-ax mill is used probably more than any other type of home-made mill. The stock on large ranches is watered by using such mills for pumping purposes. Where one has not sufficient power, two are used. The cheapness of these mills is a consideration; very seldom do they cost more than $1.50 outside of what can be

FIG. 204—MOCK TURBINE WINDMILL

picked up around the farm. The axle can be made of a pole smoothed up at the ends for bearings, or a short rod can be driven in at each end. The tower can be made of three or four poles and the sails of pole cross pieces and old boxes. One of these mills 10 feet in diameter will pump water for 75 head of cattle. Near Verdon, Nebraska, a farmer uses one of these mills in the summer to pump water for irrigation, and in the winter for sawing wood.

Turbine windmills.—The term windmills as it is commonly used refers only to the American type of shop-made mills. They may be classified by the form of the wheel and the method of governing.

1. Sectional wheel with centrifugal governor and independent rudder (Fig. 206).
2. The solid-wheel mill with side-vane governor and independent rudder (Fig. 207).
3. Solid wheel with single rudder. Regulation depends upon the fact that the wheel tends to go in the direction it turns. To aid in governing, the rudder is often placed outside of the center line of wheel shaft (Fig. 208).

FIG. 205—RECONSTRUCTED TURBINE

4. Solid or sectional wheel with no rudder back of tower, the pressure of the wind being depended upon to keep the mill square with the direction of the wind. Regulation is accomplished with a centrifugal governor (Fig. 209).

The use of the windmill.—The windmill receives its power from the kinetic energy of the moving atmosphere. Since this is supplied without cost, the power furnished by a windmill must be very cheap, the entire cost being that of interest on the cost of plant, depreciation and maintenance. Where power is wanted in small units the windmill is a very desirable motor, provided—

1. The nature of the work is such as to permit of a suspension during a calm, as pumping water and grinding feed.
2. Some form of power storage may be used.

Wind wheels.—T. O. Perry built a frame on the end of a sweep which revolved in an enclosed room in such a manner that he could fasten different wheels on it without making any change in the mechanism. By this means he was able to make very exhaustive experiments without being retarded by atmospheric conditions. He made

FIG. 206—SECTIONAL WHEEL WITH CENTRIFUGAL GOVERNOR AND INDEPENDENT RUDDER

tests with over 60 different forms of wheels, and it was the result of these experiments which brought out the steel wheel. From Mr. Perry we learn that in wood wheels the best **angle of weather** is about 30°, and that there should be a space of about one-eighth the width of the sail between the sails. By **angle of weather** is meant the angle made by the blade and the plane normal or perpendicular to the direction of the wind. With the tower in

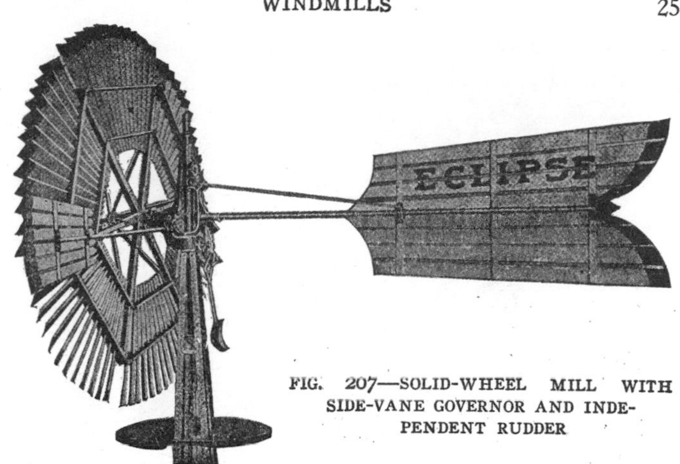

FIG. 207—SOLID-WHEEL MILL WITH SIDE-VANE GOVERNOR AND INDEPENDENT RUDDER

FIG. 208—SOLID WHEEL WITH SINGLE RUDDER

FIG. 209.—SECTIONAL WHEEL WITH NO RUDDER

front of the wheel there is a loss of efficiency of about 14 per cent; with it behind the wheel there is a loss of only about 7 per cent.

Regulation.—Wind wheels of this country are made to regulate themselves automatically, and by this means of regulation they do not attain a very high rate of speed, nearly all of them cutting themselves out when the wind has reached a velocity of about 25 miles an hour. This is principally due to the fact that our mills are generally made for pumping purposes and the pumps do not work well when the number of strokes becomes too great. It is for this reason that the direct-connected wooden wheels do not give as much power as the back-geared steel wheels. As a result of the wind wheels being thrown partially out of gear when the wind velocity is only about 25 miles an hour, many wheels are kept from doing the amount of work which they might be able to do. Any mill should stand a velocity of at least 40 miles an hour. It is understood that as the wind increases, the strain on the working parts decreases. For any given velocity of wind the speed of the wheel should not change, but the load should be so arranged that the work can be done to suit the wind.

The **efficiency** of a wind wheel is very greatly affected by the diameter. This is due to the fact that wind is not the same in any two places on the wheel. The smaller the wheel, the greater efficiency. Experiments were attempted to get the efficiency of a 22-foot wheel, but because the wind did not blow at the same velocity on any two parts of the wheel they were given up.

Gearing.—At one time the wind wheel seemed to be the most vital part of a windmill, but from the results of tests and experiments this belief has been obliterated, and now the vital part seems to be the gearing. On all the old standard makes the gearing seems to be as good as ever, even if the mills have run for several years. However, on the new designs, and this is mostly the steel mill, the gears are wearing out. The fault lies with no one but the manufacturers. Competition has been so strong that they have reduced the cost of manufacture at the expense of wearing parts. For this reason the steel wheel, which is far the more powerful, is going out of use in some localities, and the old makes of wooden wheels are coming back.

In direct-connected mills the main bearings should be long and so placed that they will carry the wheels in good shape, and the guide should be heavy and designed so that it can be lubricated easily. The bumper spring should be well placed, not too close in, so that as the wheel is thrown out of the wind there is not too much jar. Rubber should never be used for this spring, as the continual use and exposure to the weather will cause it to harden or flatten so that it is of no use. Generally weights are better to hold the wheel in the wind than springs.

In support of back or forward geared mills there is not much more to say than has been said about direct connected. The most vital parts of these mills other than named above are the gearings. They must be well set and well designed so that when they wear there is not a very great chance for them to slip.

Power of windmills.—Probably there is no other prime mover which has so many variables depending upon it as the windmill, when we undertake to compute the power by mathematical means. It is also hard to distinguish between the greatest and the least of these variables, so the author gives them promiscuously. Variable velocity of wind; velocity greater on one side of wheel than on the other; angle of weather of the sails; thickness of sails; width of sails; number of sails; length of sails; obstruction of tower either behind or in front of wheel; diameter of wheel; velocity of sails; variation of load, and location and height of tower. In all the tests of windmills which have been carefully and completely carried out it is shown that as the wind velocity increases or decreases the load should increase or decrease accordingly; as the velocity of the wheels increases, the angle of weather should decrease, and vice versa. Wide sails give more power and a greater efficiency than narrow sails.

A. R. Wolff gives the following table as results for wood-wheel mills:

Size of Wheel	R. P. M. of Wheel	Gallons of Water raised per Minute to an Elevation of						H. P. Developed
		25′	50′	75′	100′	150′	200′	
10′	60-65	19.2	9.6	6.6	4.7			0.12
12′	55-60	33.9	17.9	11.8	8.5	5.7		0.21
14′	50-55	45.1	22.6	15.3	11.2	7.8	5.0	0.28
16′	45-50	64.6	31.6	19.5	16.1	9.8	8.0	0.41
18′	40-45	97.7	52.2	32.5	24.4	17.5	12.2	0.61
20′	35-40	124.9	63.7	40.8	31.2	19.3	15.9	0.78
25′	30-35	212.4	107.0	71.6	49.7	37.3	26.7	1.34

The above table is given where the wind velocity is such that the mill makes the number of revolutions a minute given; of course, if the velocity increases, the R.P.M. will increase likewise and consequently the power.

Smeaton drew from his experiments that the power increases as the cube of the wind velocity and as the square of the diameter of the wheel. Murphy did not check this result, but found that the power increases as the squares of the velocity and as about 1.25 of the diameter of the wheel. This latter conclusion is probably the more reliable, as the instruments which Smeaton used were more crude than those of Murphy. The former determined the velocity of the wind by taking the time which it would take a feather to travel from one point to another as the velocity. The latter used a Thompson anemometer.

Tests of mills.—The following tests were made by E. C. Murphy to determine what windmills actually did in the field, also to see whether mills in practice carried out the rules made by previous experimenters. Perry found by his experiments in a closed room that the power of a wheel increases as the cube of the velocity, while Murphy found that it varied from this.

It will be noticed from the following table that some steel wheels as well as wooden gave much more power

Name	Kind	Diameter in Feet	Number Sails	Angle of Weather	Velocity of Wind in Miles per Hour	Horse Power
Monitor	Wood	12	96	34°	20	.357
Challenge	"	14	102	39°	20	.420
Irrigator	"	16	10	39°	20	.400
Althouse	"	16	130	32°	20	.600
Halliday*	"	22.5	144-100	25°	20	.890
Aermotor	Steel	12	18	31°	20	1.050
Ideal	"	12	21	32°	20	.606
Junior Ideal	"	14	24	29°	20	.610
Perkins	"	14	32	31°	20	.609
Aermotor	"	16	18	30°	20	1.530

*This wheel was made up of two concentric circles of sails, the outer having 144 sails and the inner 100.

than others. This is due to workmanship and angle of weather.

It is very clearly shown that the steel wheel is much more powerful than the wooden.

Another important factor noticed from the above table is that the 16-foot mill develops only about 50 per cent more power than the 12-foot. Taking the shipping weights of the 12-foot and 16-foot mills with 50-foot steel towers, it is found that they are about 2,000 pounds and 4,200 pounds, respectively, and since a 16-foot mill is much more liable to be damaged by a storm than a 12-foot, it is better in a great many cases to put up two 12-foot mills instead of one 16-foot.

Mr. Murphy made tests of a Little Jumbo mill 7¾ feet in diameter with eight sails, each 11×16 feet, and found that in a 20-mile wind he got 0.082 H.P. and in a 25-mile wind he got 0.100 H.P. He also made tests of a Little Giant mill and by computation found that the latter mill, having the same dimensions as the former, would start in a slower wind and when at full speed would develop about 2.5 times as much power. Other advantages of this mill over the former are that it is always in the wind and is much less liable to be injured by storms.

By a comparison of tables from different manufacturers of windmills the following table has been compiled of the size of steel windmills required for various lifts and size of cylinder. Although it cannot be said that the table is accurate, it conforms very closely to the general practice.

Size of Mill	Velocity of Wind	Size of Cylinder in Inches	Height of Lift	Size of Cylinder	Height of Lift	Size of Cylinder	Height of Lift	Size of Cylinder	Height of Lift	Size of Cylinder	Height of Lift
6	15	2″	100′	3″	50′	4″	25′				
8	15	2	100′	2½″	100′	3″	75′	4″	35′		
10	15	2″	300′	2½″	200′	3″	150′	4″	70′		
12	15	2	500′	2½″	375′	3″	250′	4″	125′		
16	15	2½″	800′	3″	500′	3½″	400′	4″	300′	5″	200′
20	15	3½′	800′	4½″	500′	5″	400′	7″	200′	8″	135′

The above table is for mills back-geared about 10 to 3. Since wood-wheel mills are generally direct-stroke, they require a much larger wheel to accomplish the same work as the steel wheels.

Towers.—The Hollanders built their towers in the form of a building which either had a revolving roof or the tower itself revolved. Within the tower they kept mills and grain. Often to-day we see the towers of American mills housed in a similar way, with the exception that they do not revolve. This is not an economical way of providing room, for it requires much more material in the construction than a low building does to withstand the excessive wind pressure which it receives.

Since the top of the tower vibrates greatly, the tower needs to be very stiff. Probably a wood tower is stiffer

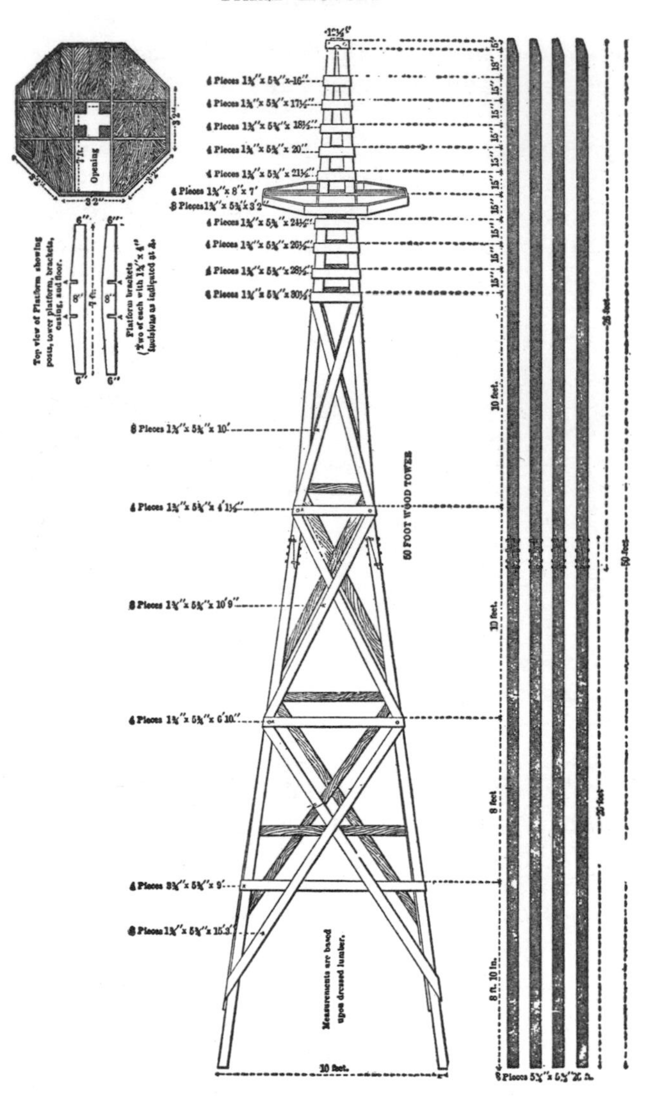

FIG. 210—DIMENSIONS FOR 50-FOOT TOWER

than steel when new, but owing to the variation in wind velocity and direction it is only a short time before the continual vibration has worked the tower loose at all joints and splices. At every joint in the wood tower there is a chance for the rain to run in and cause decay. Therefore as an offset to the greater rigidity of the wood tower one must consider the time for tightening bolts, labor for painting, and money for replacing the tower every few years.

Steel towers, as a rule, are not as rigid when new as the wood, but they do not present as great a surface to the wind as the latter, and since all parts are metal there is no chance for a loosening of the joints. The steel tower not only saves all of the labor and expense required to keep the wooden tower in repair, but it is practically indestructible.

In a cyclone the steel tower will often become twisted before the wooden one will be broken. However, the latter will generally become so racked and splintered that it cannot be repaired.

Anchor posts can be made by setting strong fence posts in the ground their full length and nailing some strips across them to hold beneath the earth; but a better method is to insert an angle iron in a concrete base, which will support the tower posts. The dimensions of the base should be about 18 × 18 inches × 4 feet for small mills, and proportionally larger for large mills.

Erecting mills.—Windmills over 60 feet high should be assembled piece by piece, but low towers can be assembled on the ground, including windmill head, sails, and vanes, then raised in a manner similar to Fig. 211. After the tower has been raised it should be examined and all braces and stays given the same tension and all nuts tightened. It is also well before the pump rod is put in place to drop a plumb bob from the center of the top of the tower to the intersection of cords stretched diagonally from the corners of the tower at

FIG. 211—RAISING A TOWER

the base. If the plumb bob does not fall on this intersection, either the braces do not have equal tension or the anchor posts are not level.

Economic considerations of windmills.—Many manufacturers claim much more power than the windmills really develop. This erroneous claim is probably due to the fact that early experimenters worked with small wheels and figured the power of larger ones from the law of cubes, which does not seem to hold true in actual practice. It is wrong to say that a good 12-foot steel mill will furnish 1 H.P. in a 20-mile wind and that a good 16-foot mill will furnish 1.5 H.P.

The economic value of a windmill depends upon its first cost, its cost of repairs, and its power. The competition in manufacture at present is so great that often the initial cost is kept down at the expense of the other two.

A mill should have as few moving parts as possible. The power of a mill is so small that if there is much to retard its action there will be very little power left for use.

In power mills very often the shafting is much heavier than need be. This is probably due to the fact that the mill was designed for much more power than it will actually develop. Often poor workmanship in manufacture as well as in erection is the cause of so many mills having such small power.

Trees, buildings, and embankments cause the wind velocity to be so variable that for good work it is desirable that the wind wheel be placed at least 30 feet above all obstructions. This would cause the towers to be at least 60 or 70 feet high. It is better to put a small wheel on a high tower than a large wheel on a low tower. An 8-foot wheel on a 70-foot tower will probably do more work in a given length of time than a 12-foot wheel on a 30-foot tower.

The pumping mill is ordinarily constructed so the work is nearly all done on the up stroke. This is hard on the mill, as it produces a very jerky motion and excessive strain on the working parts. By placing a heavy weight on one end of a lever and connecting the plunger rod to the other this strain is reduced, since when the plunger rod goes down it raises the weight, and when it comes up, lifting the pump valve and water, the weight goes down and thus assists the mill.

How the wind may be utilized.—In a country where there is such an abundant supply of wind as in the Central and Western States there is no doubt that a windmill is the cheapest and most feasible power for the farmer. In certain localities water power is a great opponent of the wind, but it has the disadvantage to the farmer of being in the wrong location, causing water rights to be looked after and dams to be kept in repair, while in utilizing the wind all that is required is some simple device which will turn wind pressure into work.

The windmill without doubt is the best machine for this, but since we cannot depend on the wind at all hours of the day, we must devise some scheme whereby we can store the work when the wind blows so that we may use it when there is no wind. For this means four ways come to mind: One is to connect a dynamo to the mill and store the electricity in storage batteries. This is not a feasible plan at present, since the expense of storage batteries and the cost of repairs is too great. Another plan is to run an air compressor by means of the wind and then use the compressed air for power purposes. This again is not satisfactory owing to the cost of keeping air machines in repair and also of conveying the air. Another scheme, and probably the best, is to pump water into a tank on a tower, and then let this water which has been stored up during the time of wind run down through a water motor and from thence to the yards, or, if there is more water than is desired for the stock and house use, run it into another tank below the tower and then pump it back. Another scheme which is similar to that named last is to pump the water into a pressure tank in the cellar and then let it pass out the same as in the tank on the tower. By this latter scheme the expense of the tower and the danger of freezing are obviated, but a more expensive tank and also an air pump are added.

Power mills.—The same discussion, which has been given more especially to pump mills, will apply to power mills. As a rule, power mills are larger than pump mills, and require more skill in keeping the bearings in repair. Care should be taken in erecting power mills that the shaft is in perfect alignment. A great deal of power can be lost by not having the shaft running in a perfect line.

CHAPTER III

STEAM BOILERS

Principle.—A kettle over the fire filled with water is a boiler of small proportions. When fuel is burned beneath the kettle heat is transferred to the metal of the kettle and from the metal to the water at the bottom. Thus the water in direct contact with the bottom is heated, and, since warm water is lighter than cold, the warmer water rises to the top and the cold settles in its place. In physics this action of the water rising and falling in the kettle, conveying the heat from one part to another, is known as convection. In the steam boiler it is known as circulation. When sufficient heat has been transferred to the water to raise the temperature to 212° F. it will commence to boil and throw off steam.

The reason why the water had to be heated to 212° before the particles of water would be thrown off as steam was because the atmosphere, having a pressure of 14.7 pounds to each square inch, pressed upon it so hard that the steam could not be thrown off until this temperature had been reached. If the kettle were up on a mountain where the atmospheric pressure is not nearly as great, steam would have been thrown off at a lower temperature.

The same process which takes place in a steam boiler also takes place in a kettle, only under less economical conditions. A fire is maintained within the furnace of the boiler and the heat is transferred to the metal of the boiler shell and tubes, thence to the water, which is converted into steam. The water of a low-pressure boiler, i.e., one which carries a pressure of only about 5 pounds gauge, is heated to only about 228° when steam is given off, while in a high-pressure boiler which carries about 200 pounds gauge pressure it has to be heated to about 385°.

The first boilers were simply large cylindrical shells. They did the work required of them, but were very in-

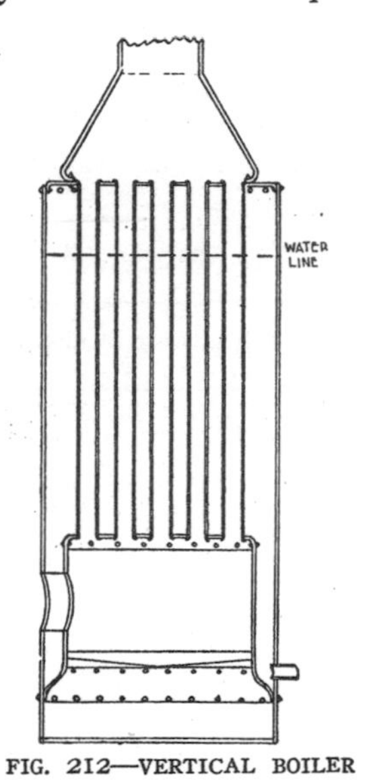

FIG. 212—VERTICAL BOILER

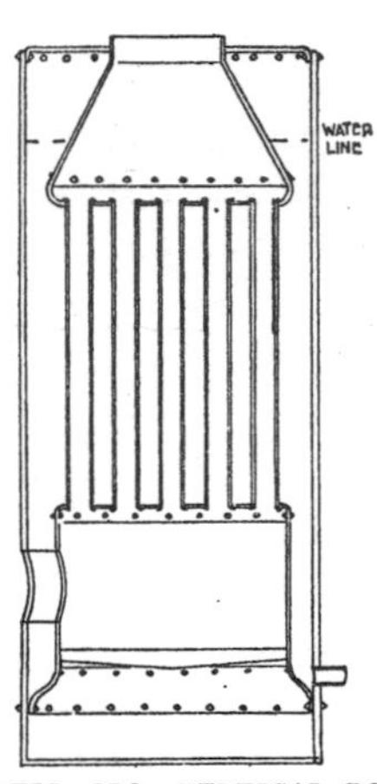

FIG. 213—VERTICAL BOILER WITH SUBMERGED FLUES

efficient. The next was merely a shell with one tube or flue, as it is often called. Multitubular, return tubular, internally fired, water-tube, sectional boilers, etc., have come in in succession until we have the present-day types.

Classification.—Steam boilers may be classified according to their form and use. Thus we have locomotive, marine, portable, semi-portable, and stationary boilers, according to use; and according to form we have horizontal and vertical boilers. Further, the horizontal class may be subdivided into internally and externally fired, shell, return-flue, fire-tube and water-tube boilers. For

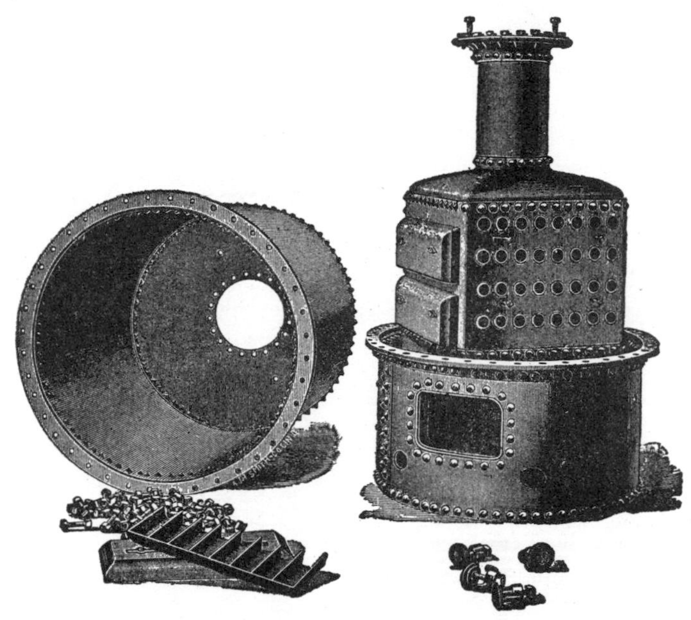

FIG. 214—WATER-TUBE VERTICAL BOILER

rural use the marine type is very seldom used, and the sectional only in rare cases.

Vertical boilers.—Boilers of this type (Fig. 212) are not very economical. They require little floor space and are easily installed. In construction they consist of a vertical shell, in the lower end of which are the fire box and ash pit; extending up from the furnace and reaching the top are the fire flues.

Since the shell of the fire box is under external pressure, it must be stayed to avoid collapsing. The blow-off cock and frequent hand holes are near the base for

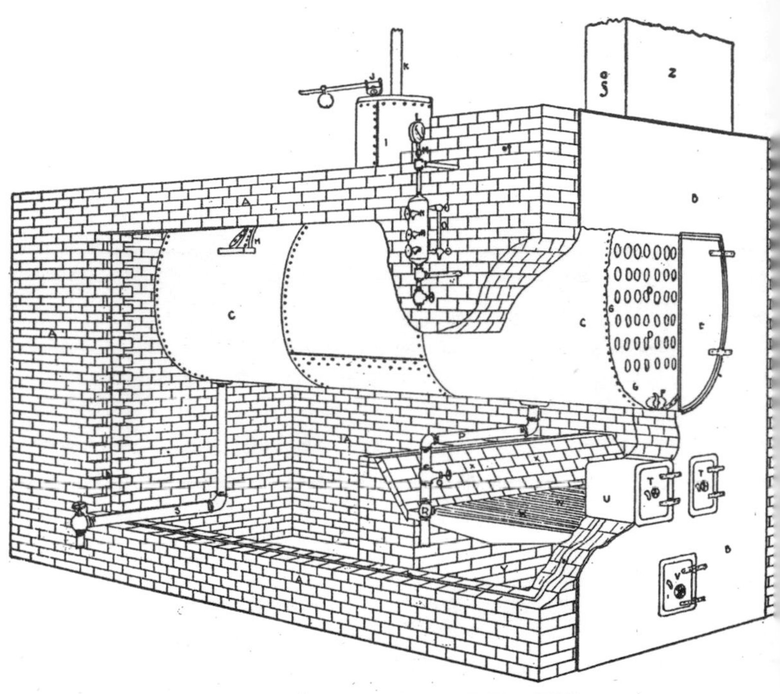

FIG. 215—EXTERNALLY FIRED BOILER

AA, boiler setting; *BB*, boiler front; *CC*, boiler shell; *DD*, flues; *E*, flue door; *F*, handhole; *G*, flue sheet; *H*, bracket; *I*, steam dome; *J*, safety valve; *K*, steam pipe; *L*, steam gauge; *M*, steam gauge syphon; *NN*, try cocks; *O*, water glass; *PS*, blow-off pipes; *Q*, blow-off valve; *TT*, fire door; *U*, fire door lining; *V*, ash door; *W*, grates; *X*, bridge wall; *Y*, ash pit; *Z*, britchen; *A*, damper.

convenient cleaning. A water glass and try cocks are near the top. Heating surface in this type of boiler consists of the fire box and the fire tubes up to the water line; as the water does not completely cover the tubes, the upper part forms a superheater.

When the exhaust steam is released into the stack, the tubes have a tendency to leak. To avoid this, some manufacturers sink the tube sheet below the water level (Fig. 213). This form reduces the superheating surface, and moreover, since the conical smoke chamber is subjected to internal pressure, it is likely to be weak. Fig. 214 is a special type of vertical boiler in which are water tubes laid up in courses. The boiler shell can be removed from the caisson of tubes so that all parts are accessible for cleaning and repairing.

Externally fired boilers (Fig. 215) are generally of the cylindrical tubular type and can be used for stationary work only. These are probably the most simple as well as most easily handled and kept in repair of all, but they are very bulky, requiring a great amount of floor space. The furnace for such boilers is a part of the setting and is made under the front end. The flames surround the lower part of the shell and pass to the rear, where they enter the tubes and return to the front, thence up the stack.

When setting externally fired boilers, care should be taken that one end or the other, generally the rear, be free to move forward or backward, since the variation of temperature will cause the boiler to contract and expand enough to crack the masonry upon which it rests.

Internally fired boilers.—This class comprises several types, the locomotive type (Fig. 216), the return-flue type (Fig. 217), and the Lancashire. The first two of these types are the most used for traction or portable work, while the latter is adapted only to stationary use.

Locomotive type.—The locomotive fire-tube type was probably the first of the modern boilers to come into

FIG. 216—LOCOMOTIVE TYPE OF INTERNALLY FIRED BOILER

general use. With only a few changes, it is the same now. By referring to Fig. 218, it will be noticed that the fire box is practically built into the rear end of the boiler barrel. Extending from the rear tube sheet and through the entire length of boiler barrel are the fire tubes, which are generally about two inches in diameter. Surrounding the fire box and fire tubes is the water. This gives abundance of heating surface, also freedom of circulation. As the sides of the fire box are nearly flat, they will easily collapse under the pressure of the steam

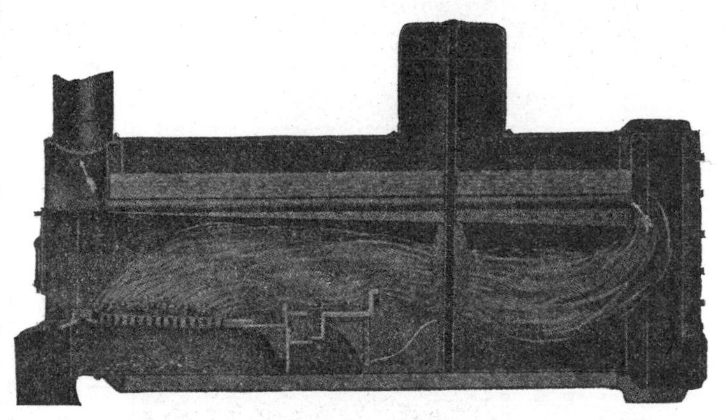

FIG. 217—RETURN-FLUE TYPE OF INTERNALLY FIRED BOILER

unless supported by stay bolts at intervals of every few inches.

The steam dome can be located anywhere, but it is generally placed about midway between front and rear ends. A pipe takes the steam from the top of the dome, carries it down through the steam space, where it is dried, then out wherever convenient.

Generally the blow-off is at the bottom and in front of the fire box. The water glass is placed about on a level with the crown sheet, since this is the place where the water must not get low.

Round-bottom types.—The principal variation from the original type of this class of boilers is in the design of the rear or furnace end. The common practice is to have the water pass completely around the fire box, including the under side. Such boilers are generally known as the round- or enclosed-bottom type (Fig. 218). As a rule, the draft can enter at front or rear of the fire box. This method of draft frequently aids the fireman in firing up, for when there is but one ash door the direction of the wind may be such as to blow away from the door, retarding the draft.

The open-bottom type (Fig. 219) is so constructed that ash pan and grates can be removed and a complete new fire-box lining put in. The draft can enter at either end of the fire box. There is not as free circulation in this type as in the round-bottom boilers, providing the latter are kept clean.

When a portable boiler of the locomotive type is setting with the front end low, unless there is an abundance of water, the crown sheet will be exposed and, if not attended to at once, will become overheated and collapse. To aid in avoiding this, some manufacturers are making the rear end of the crown sheet (Fig. 220) lower than the front. This mode of construction reduces the size of the rear end of the fire box to a certain degree, but it is done where the space is not essential. Fig. 220 also shows a device which further aids in protecting the crown sheet by displacing the water in the front end of the boiler.

Return-flue boilers of the internally fired type have one main flue, which carries the gases from the fire box through the boiler to the front end. Here they

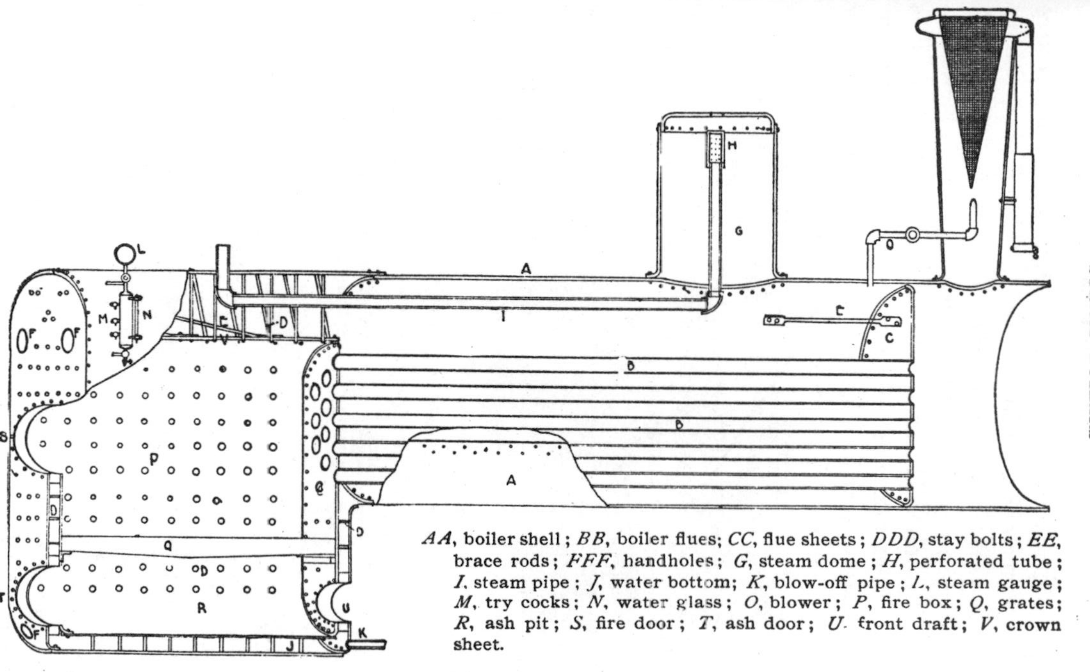

AA, boiler shell; *BB*, boiler flues; *CC*, flue sheets; *DDD*, stay bolts; *EE*, brace rods; *FFF*, handholes; *G*, steam dome; *H*, perforated tube; *I*, steam pipe; *J*, water bottom; *K*, blow-off pipe; *L*, steam gauge; *M*, try cocks; *N*, water glass; *O*, blower; *P*, fire box; *Q*, grates; *R*, ash pit; *S*, fire door; *T*, ash door; *U*, front draft; *V*, crown sheet.

FIG. 218—SECTIONAL VIEW OF LOCOMOTIVE TYPE OF INTERNALLY FIRED BOILER

are divided and enter several smaller flues, then return to the rear end and pass up the stack. This, without doubt, is a very economical type.

By referring to Fig. 221, which is an end view of a return-flue boiler, it will be noticed that the smaller tubes are above the main flue. By this arrangement the smaller and cooler parts will become exposed first, thus giving

FIG. 219—OPEN-BOTTOM FIRE-BOX BOILER

the engineer a chance to save the boiler from collapse or explosion.

Wood and cob burners.—Most boilers upon the market have interchangeable grates so that by placing a grate with smaller openings in place of the coarser one for coal, wood and cobs may be burned.

Since the most economical firing can be accomplished by refraining from poking the fire on top, a great many factories are making a rocker grate (Fig. 222), which is

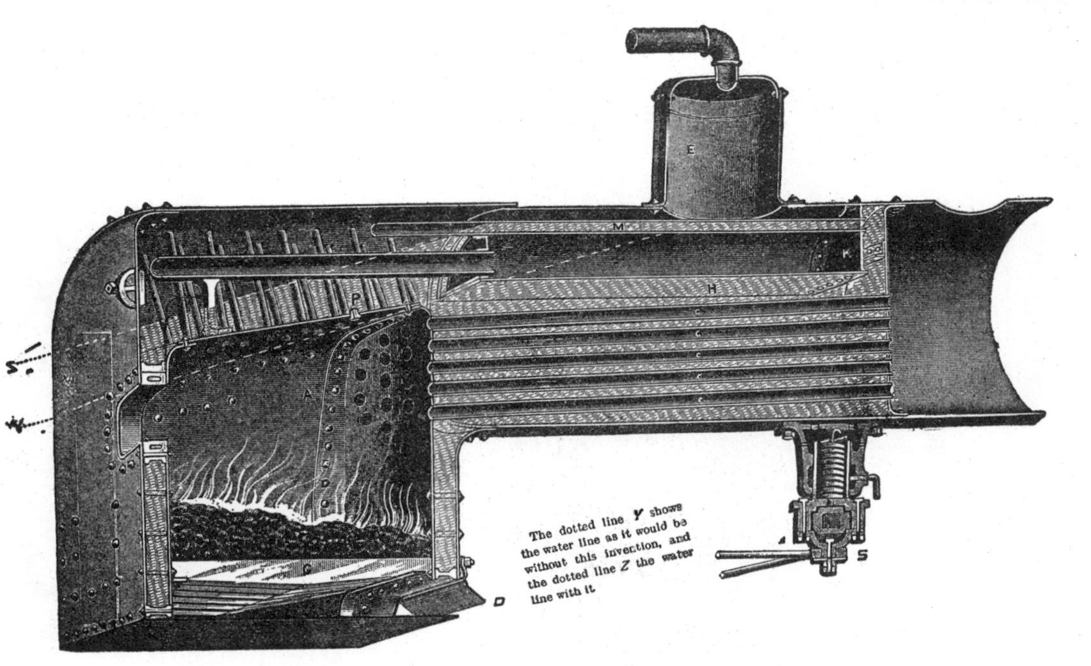

FIG. 220—BOILER WITH REAR END OF CROWN SHEET LOWER THAN FRONT END

FIG. 221.—END VIEW OF RETURN-FLUE BOILER

worked by a lever in such a manner that all fine ashes will drop through.

Straw burners.—For burning straw there must be special arrangements within the fire box. The fuel is light and generally chaffy, and as a result flashes up very quickly, and unless prevented will be carried by the draft some distance through the tubes before it is all aflame. Not only this, but straw must be burned rapidly in order to produce heat enough to make steam as fast as needed. To handle straw under these conditions, the return-flue boilers are generally constructed similar to the type shown in Fig. 223: *a* is an extended fire box with a drop-hinge door; *b* is the upper grate; and *c* is the lower grate, where as much of the straw as is not burned in the upper grate, or as it falls from it, is consumed; *d d* are deflectors which hold the flames next to the upper side of the flue.

FIG. 222.—ROCKER GRATES

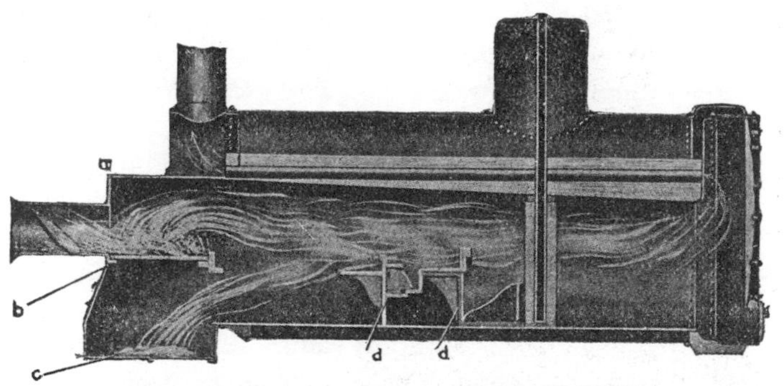

FIG. 223.—STRAW BURNER RETURN-FLUE BOILER

Direct-flue boilers, (Fig. 224), can be more easily changed from coal burners to straw burners. This is generally done by adding a feeding tube with an enclosed drop-hinge door, by removing the grates and inserting a dead plate with short grates in front of it, and by placing a deflecting arch composed of firebrick in the fire box.

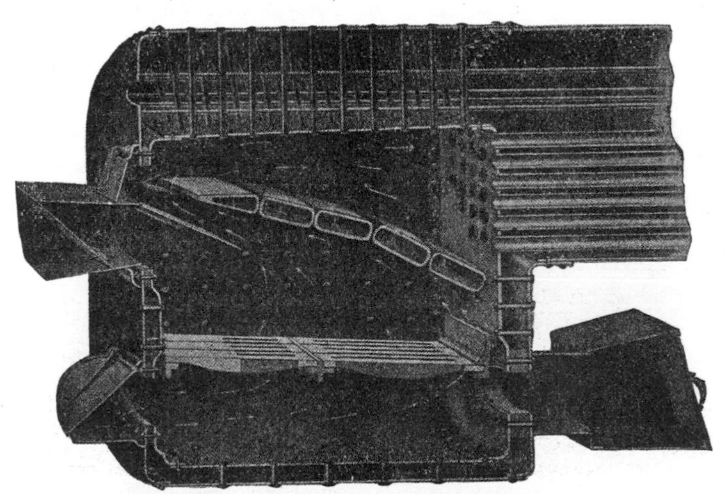

FIG. 224.—STRAW BURNER DIRECT-FLUE BOILER

By means of the shorter grates the draft opening is reduced, and by the aid of the deflector a combustion chamber is produced where all of the light particles are consumed and the gases are heated to an incandescent state before entering the tubes. The direction of draft in this type is nearly always toward the straw, thus causing the heat as it passes the unburned straw to prepare it for better combustion.

BOILER ACCESSORIES

FIG. 225—SYPHON FOR FILLING SUPPLY TANK

Supply tank.—Boilers used for traction purposes require a small supply tank to which the boiler pump or the injector is connected. This tank is generally placed in some position where it is convenient, yet out of the way.

Siphon or ejector—When the supply tank is placed so high that it cannot be filled from a stock tank or other similar source, a siphon (Fig. 225) is generally used. The construction of this is such that a jet of steam is passed into a water pipe leading from the tank or cistern to the supply tank. As the steam comes in contact with the water it is condensed; this produces a vacuum such that the water rushes in to fill, and the inertia due to the velocity of the steam sends it along into the supply tank.

Care must be taken in regard to the amount of steam used, since if too much steam be used the water will become so warm that the feed pump or injector will not work.

Feed pumps.—There are three types of pumps now in use: the crosshead pump, the independent direct-

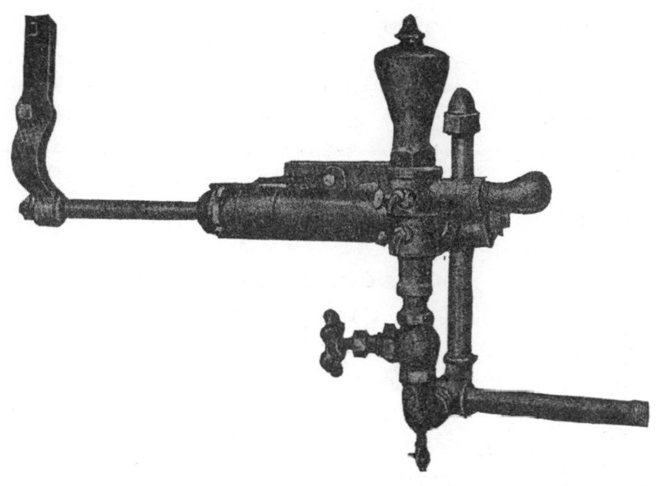

FIG. 226—CROSSHEAD PUMP

FIG. 227—INDEPENDENT DIRECT-ACTING PUMP

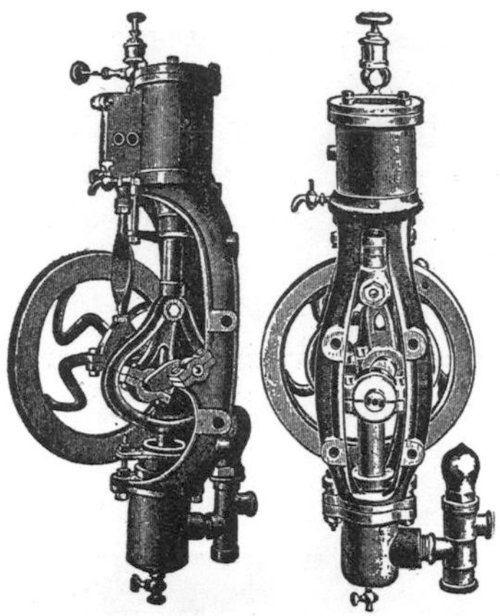

FIG. 228—INDEPENDENT PLUNGER PUMP

acting (Marsh) pump, and the independent plunger pump (Figs. 226, 227, and 228). The crosshead type is the simplest and most economical, but can be run only when the engine is running. The Marsh independent pump is simple and economical, but the action of its steam valve is delicate and should be molested only by an expert. The independent plunger pump is very satisfactory in that it can be run at any time and by any one. The initial cost of this is more than that of other types.

The injector is probably the most generally used means of feeding boilers. It was invented in 1858 by M. Giffard, and large numbers of the same types are still made. The action of the injector will be understood by referring to the sketch (Fig. 229). Steam is taken from the boiler and passes through the nozzle *A* to the injector; the amount of steam is

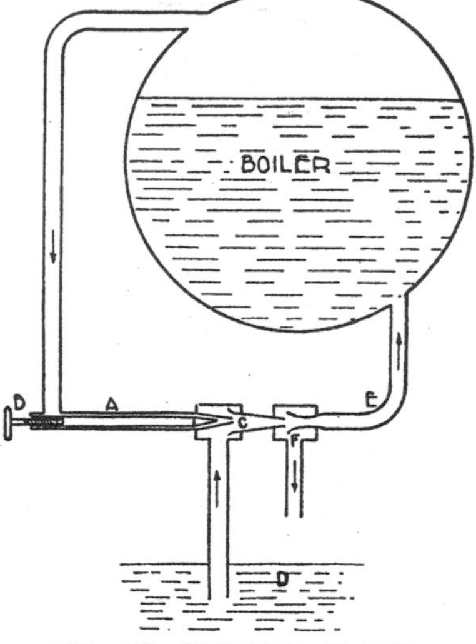

FIG. 229—PRINCIPLE OF THE INJECTOR

regulated by the valve *B*. In the tube *C* the steam is combined with the slowly moving water, which is drawn up from the tank *D*. The swiftly flowing steam puts sufficient momentum into the water to carry it into the boiler. The delivery tube *E* has a break in it at *F* where the surplus steam or water can overflow.

An injector should be chosen with reference to the special work required of it. Some will lift water, others will not. Some will start under low-pressure steam and

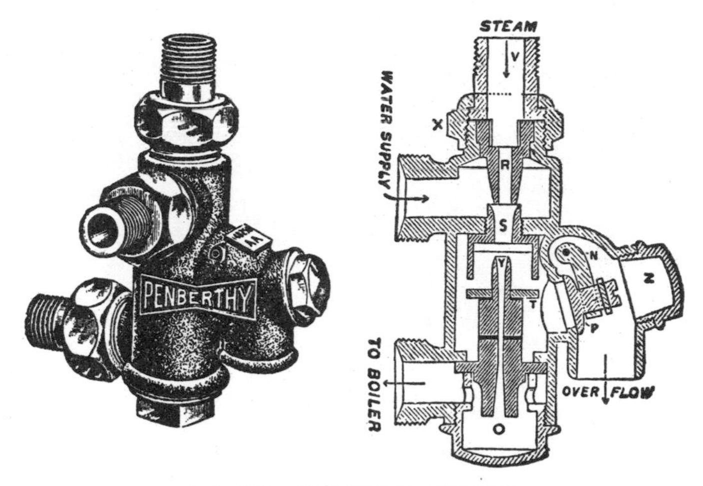

FIG. 230—COMMERCIAL INJECTOR

refuse to act under high, while with others the reverse is true. There are also injectors which will operate with exhaust steam. Such an injector is not essential, since the efficiency of one of high pressure is practically 100 per cent.

Locomotives are equipped with self-starting injectors. Every traction engine should be equipped with two systems of boiler feeds. Some have two injectors, while some have two pumps, but the most common method is a pump and injector.

Feed-water heaters.—The sudden change in temperature of boilers puts them under a great deal of strain. One of the principal reasons for this change in temperature is the admitting of cold feed water. This water may be easily heated by passing the exhaust steam through it. There are two methods of such heating: one is to allow the exhaust steam to mingle with the water, thus being condensed and carried back to the boiler, and the other is to pass the feed water through pipes surrounded by steam. By the former method the steam is returned

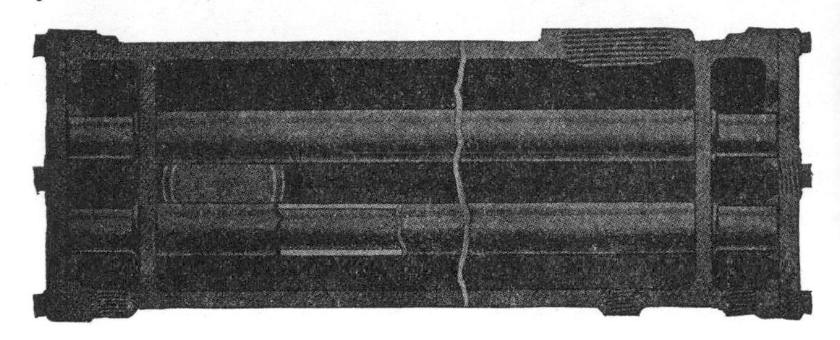

FIG. 231—FEED-WATER HEATER

to the boiler, and unless a filter is used all the cylinder oil is carried into the boiler, to which it is detrimental. In the latter case the steam does not return to the boiler, but is sent up the stack, thus producing a forced draft. Fig. 231 shows a heater of this type.

As pumps and injectors will not operate with hot water, and since the water from a heater is nearly as hot as the exhaust steam, the heater must be located between pump and boiler.

Water columns.—The purpose of the water column is to support the gauge glass and try cocks; it is used only in stationary boilers. The water column should be located so that the center of the column will come to the point where the level of the water should be above the tubes, or crown sheet. The column is generally of a casting about 3½ inches in diameter and 15 inches long. Into this casting are secured the try cocks and water glass. Some builders connect the steam gauge to the upper end.

By referring to Fig. 232 it will be noticed that the lower end of the glass, the lower try cock, and the crown sheet are on a level with each other, hence when the water is out of sight in the glass and also will not flow from the try cock the crown sheet is exposed. The water should be kept about in the middle of the glass, and likewise even with the center try cock. It should not be above the upper try cock, or there will be trouble from wet steam.

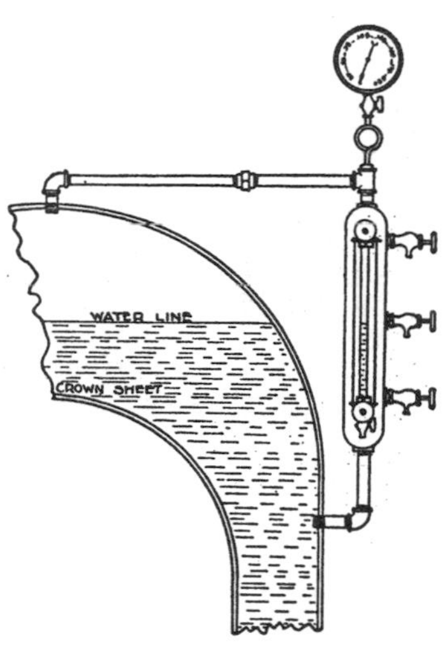

FIG. 232—WATER COLUMN, GAUGE GLASS, TRY COCKS AND STEAM GAUGE

Steam gauge.—The mechanism of a steam gauge (Fig. 233) usually consists of a thin tube bent in a circle. One end of the tube is connected to the boiler, and the other, by means of a link, to a small pinion which works a needle indicator. Air is kept in the tube by means of the siphon, and a cylinder of water lies between the air and the boiler. When there is zero pressure in the boiler the needle should set at 0. As pressure begins to rise in the

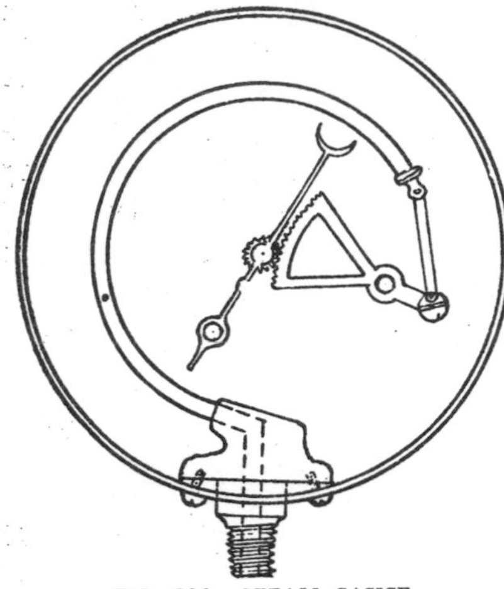

FIG. 233—STEAM GAUGE

boiler the air will tend to straighten the tube, and hence the tube acts upon the needle. If it is found by comparison with another gauge that the needle does not indicate the actual steam pressure it can be regulated by sliding the link up or down in the slot at the end of the pinion, thus changing the throw of the needle.

Fusible plug. — As a safeguard against low water a fusible plug is put in the boiler. In fire-box boilers it is placed in the crown sheet directly over the fire, and in return-flue boilers it is placed in the back end just above the upper row of flues. The plug is generally made of brass about one inch in diameter and with a tapered hole bored through its center (Fig. 254). The tapered hole is filled with some metal, generally Banca tin, which will fuse at a low temperature, so that when the water has become so low that the metal melts and runs out the steam will flow through the opening and put out the fire.

Safety or pop valve.—It is essential that in every boiler there be a safety valve so that the steam may be released before too high pressure has been reached. There are two distinct types of these valves, the ball and lever valve and the spring pop valve. The former (Fig. 235) is

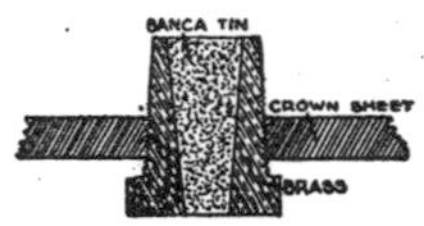

FIG. 234—FUSIBLE PLUG

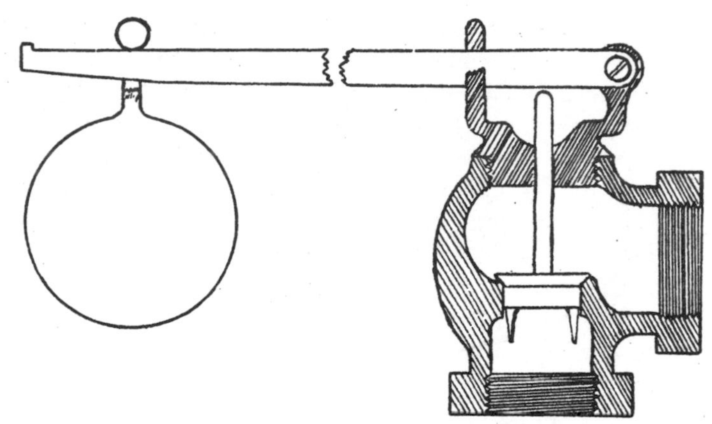

FIG. 235—BALL AND LEVER SAFETY VALVE

the least expensive, also the less reliable. It is generally used upon stationary boilers. To increase the pressure in the boiler before it blows off, the ball must be moved farther out on the lever, and inversely to decrease the pressure. The ball should be set at the proper point to blow off at the desired pressure, and then the lever marked so that the point can be seen distinctly.

Spring safety valves are generally used on traction engines and the better class of boilers. They are more reliable and also act much more quickly. If properly constructed they will allow the pressure to fall about 5 pounds before closing, while the ball and lever type only falls to a trifle less than the blow-off pressure. By referring to Fig. 236

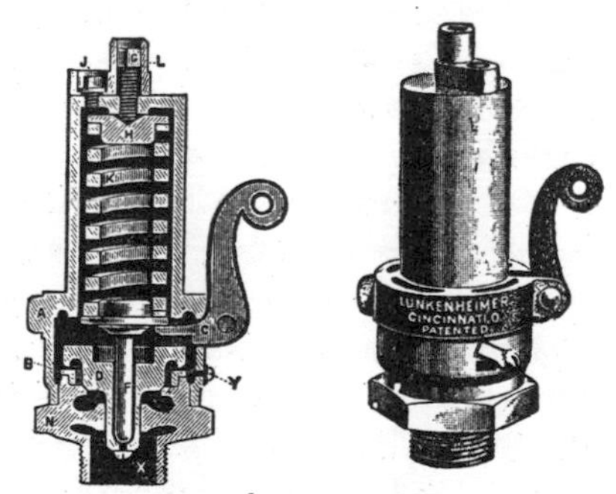

FIG. 236—POP VALVE

it will be noticed that there is a groove *B* in the valve such that when the valve starts to open, the steam rushes into it, thus increasing the area of the valve and causing it to open more quickly and remain open longer. To increase the pressure at blow-off, screw down on the pin *G*; to lower the pressure, screw up on the pin *G*. Care must be taken not to tighten the spring down too far, or it will not allow the valve to lift off its seat.

Blower and exhaust nozzle.—In all traction engines there must be some method of increasing the draft. The most simple method and the one universally used is the blower when the engine is not running, and the exhaust when it is.

The blower (Fig. 218) consists of a small pipe with a valve which leads from the boiler to the stack. After the pressure has reached 5 or 10 pounds the valve in this pipe is opened and a jet of steam is allowed to blow into the stack. The momentum of the steam produces a vacuum and the air rushing through the grates and coal to fill this space increases the rate of combustion. When the engine is running the exhaust steam from the heater takes the place of the blower and the latter is closed. Fig. 237 shows an exhaust nozzle which can be made to give a sharp or sluggish exhaust, as desired.

FIG. 237—EXHAUST NOZZLE

Blow-off pipe.—Wherever there is a chance for sediment of any kind to collect in a boiler there should be some means of cleaning it. This is almost always accomplished by means of a blow-off pipe and valve. In vertical boilers this is located at the lower end of the water leg. In return-flue boilers this is either at the front or the rear end, and in fire-box boilers it is beneath the fire box or in the water legs.

Spark arrester.—Where some method of forced draft is used in a boiler there is danger of sparks being carried out and causing fires. Traction engines guard against this by means of a spark arrester. This may consist of a screen which catches the sparks and allows them to fall into the stack, or it may be accomplished by turning the smoke around a sharp corner and, as the sparks are heavier than the smoke, they will be thrown out and are caught in a receptacle for that purpose. The smoke box or front end of the boiler may be long for the purpose.

BOILER CAPACITY

The **capacity** of a boiler depends upon the amount of heat generated and the proportion of that heat transferred to the water. The amount of heat generated depends upon the quantity of coal, the draft, and area of grate surface. The amount of heat transferred from the coal to the water depends upon the amount and position of the heating surface.

There is no entirely satisfactory method of stating the capacity of a boiler or its economy, but they are commonly stated as **boiler horse power** and the pounds of steam evaporated per pound of coal. This method of rating is on the assumption that the steam is all dry saturated steam and that there is no priming or superheating.

When water is carried along with steam from the boiler it is called **priming.** Very seldom is a boiler designed which does not prime at least 2 per cent, but if it primes over 3 per cent it is improperly designed. When steam passes over a hot surface after leaving the boiler it will absorb additional heat and become superheated. That part of the tubes which is above the water line in a vertical boiler is superheating surface. In other styles of boilers the steam in order to be superheated generally passes through a coil of pipe within the fire box or a furnace made purposely for it.

Steam space.—The surface for the disengagement of steam and the steam space should be of sufficient size so that there is no tendency for the water to pass off with the steam. It has been found by experiment that if the steam space has capacity to supply the engine with steam for 20 seconds, there will be no trouble with priming. To determine whether the boiler has sufficient steam space, find the volumes of the engine cylinder, less the volume of the piston, and multiply this by twice the number of revolutions that the engine makes in 20 seconds. This should be about equal to the volume of the steam space, which is the space above the water in the boiler, plus that in the dome.

Boiler horse power.—There are two common methods of approximately determining the horse power of a boiler, and a third one which is sometimes resorted to. One of the common methods is by test, and the other is by heating surface, while the third method is by grate surface.

Horse power by test.—A committee of the American Society of Mechanical Engineers has recommended that one horse power be equivalent to evaporating 30 pounds of water at 100° F. under a pressure of 70 pounds gauge. This is equivalent to 33,320 B.T.U. an hour.

Example.—If a 15 H.P. boiler evaporate 15 × 30 or 450 pounds of water in one hour with feed water at 100° and under a gauge pressure of 70 pounds, it would be doing its rated horse power. To make the test, fill the boiler to its proper level and tie a string around the glass at this point, then keep the water in the boiler at this level. If the feed water is below 100°, turn steam into it until the proper temperature has been reached. Use just steam enough to keep the pressure at 70 pounds. Weigh the feed water supply before starting, then weigh again at the close of the run. If the run has been of one hour's duration, divide the number of pounds of feed water by 30, and this will give the horse power developed. If the run has been only one-half hour, multiply by 2, then divide by 30.

Power by heating surface.—The heating surface of a boiler consists of the entire area of those parts of the surface which have fire on one side and water on the other. In the horizontal tubular boiler it is all of the shell which comes beneath the boiler arch, also the inside area of all the tubes and about two-thirds the area of the tube sheets less the area of the flues. In the vertical boilers it is the total inside area of the fire box and as much of the tubes as is below the water line.

In the fire-box boilers it is the inside area of the water legs, the crown sheet, and the flues and a portion of the tube sheets.

The common rating of boiler horse power by heating surface is 14 square feet for each horse power. This varies with the boiler, some styles requiring a little less and some a little more.

As an example, let it be desired to find the heating surface of a horizontal tubular boiler. Find the total area of the outside of shell and take about one-half of this. The brickwork covers about one-half of the shell, hence, one-half of it is all the heating surface there is in this part. Now measure and compute the inside area of one of the flues and multiply this by the number of flues. Add this surface to the heating surface of the shell and divide the sum by 14. This gives the horse power of the boiler.

Power by grate surface.—This method is not very often resorted to. In any case it can be only a rough estimate. It is generally conceded that from one-third to one-half square feet of grate surface is equivalent to one horse power.

STRENGTH OF BOILERS

Materials used.—The materials used in the construction of boilers are mild steel, wrought iron, cast iron, copper, and brass.

In order that a boiler have proper strength for the severe work required of it, sample pieces of all the materials used in its construction are selected and given a test, and those which fail to have the proper requirements are discarded. They are tested in tension, compression, and shear. (See Chap. III, Part I.)

Steel.—All present-day boilers are made up of mild-steel plates. This steel is a tough, ductile, ingot metal, with about one-quarter of 1 per cent of carbon. It should have a tensile strength of about 55,000-60,000 pounds. Sometimes a better grade of steel plate is used for the fire box and tube sheets of the boiler than for the shell. This is because flanging for riveting and the variations of temperature due to the fire require a better grade of steel.

Blue heat.—All forms of mild steel are very brittle when at a temperature corresponding to a blue heat. Plates that will bend double when cold or at a red heat will crack if bent at a blue heat.

Wrought-iron parts.—All welded rods and stays should be of wrought iron. About 35 per cent of the strength of the bar is lost because of the weld. Boiler plates made of wrought iron are considered more satisfactory than of steel, but are used only in exceptional cases because of the greater cost. Wrought-iron plates should have a tensile strength of 45,000, and bolts should have 48,000.

Rivets.—Boiler rivets are either of wrought iron or mild steel. The rods from which rivets are made should have a tensile strength of 55,000 pounds for steel and 48,000 for iron. When cold they should bend around a rod of their own diameter, and when warm bend double without a fracture. The shearing strength is about two-thirds of the tensile strength.

Cast iron is used in boilers for those parts where there are no sudden changes of temperature and where there is no great tensile strength required. Couplings, elbows, etc., are better of cast iron, for when they become set and can be removed in no other way they can be broken.

Stay bolts and stay rods.—In some parts of the boilers the flues act as stays. In horizontal tubular boilers the flues hold the ends of the shell together. In the fire box and in vertical boilers they act in the same way between the flue sheets. Wherever there are flat surfaces and no other means of supporting them, special stay bolts or braces must be put in. In nearly all boilers above the flues stay rods are used to support the ends. Around the fire box stay bolts are put in. These bolts are threaded full length, then screwed through the outer shell and through the water leg and into the fire-box lining, then they are riveted on both ends. Their size and distance apart depends upon the pressure to be carried.

Example.—If the stay bolts are 4 inches apart and the maximum pressure to be carried is 120 pounds they should be large enough to hold

$$4 \times 4 \times 120 = 1{,}920$$

pounds. If we use a factor of safety of 10—that is, make it ten times as strong as necessary to avoid accidents—it will have to be large enough at the base of the thread to hold

$$1{,}920 \times 10 = 19{,}200$$

pounds. If a wrought-iron bolt is used it would have to have

$$19{,}200 \div 48{,}000 = 0.40$$

square inches area at the base of threads. A ¾-inch bolt has about this area.

Strength of boiler shell.—To determine the tension upon one side of a boiler shell, let

p = pressure in pounds per square inch,
t = thickness in inches,
r = radius,
s = stress in pounds per square inch;

then

$$s = \frac{pr}{t}.$$

Example.—A boiler has a diameter of 3 feet, a thickness of 7/16 inch and the steam pressure is 125 pounds. How many pounds per square inch pull is there on each side?

$$s = \frac{pr}{t} = \frac{125 \times 3 \times 12}{2} \div \frac{7}{16} = 5,143$$

pounds. This is about one-tenth the tension which boiler plate will stand, hence we have a factor of safety of 10, which is greater than need be.

Riveted joints.—If a boiler shell could be made of one continuous piece, the above tension would be the safe working load, but since the steel has to be riveted and a riveted joint is not as strong as the original plate, we must consider the ratio of this strength of the whole plate. This ratio is commonly called the efficiency of a riveted joint.

There are three general ways that a riveted joint may give way:

1. By tearing the plate between the rivets.
2. By shearing the rivets.
3. By crushing the rivets or plate at the point of contact.

Since only single-riveted and double-riveted lap joints are used in small boilers, these styles will be considered only.

Single-riveted lap joint.—In the joint shown by Fig. 238, let t be the thickness of plate, d the diameter of rivet, p the distance between rivets, commonly called pitch, the tensile strength of the plate $S_t = 45{,}000$, and resistance to crushing $S_c = 90{,}000$. Assume $t = 7/16$ inch, $d = 1$ inch, and $p = 2\frac{1}{2}$ inches.

A strip of the joint equal in width to the pitch is sufficient to be considered.

1. Tearing between the two rivets.—In this case there is a strip to be torn in two, equal in width to the distance between the rivets less the diameter of the rivet, i.e., $p - d$, and it has a thickness equal to t, i.e., the strip has a cross-section of an area $(p - d)t$; this cross-section in square inches times the tensile strength will give the pull required to fracture the joint:

$$(p - d)tS_t = (2\tfrac{1}{2} - 1) \times 7/16 \times 55{,}000 = 36{,}095.$$

2. Shearing one rivet.—Since there is only one rivet in each $2\frac{1}{2}$-inch strip, we have to consider the shearing of it only.

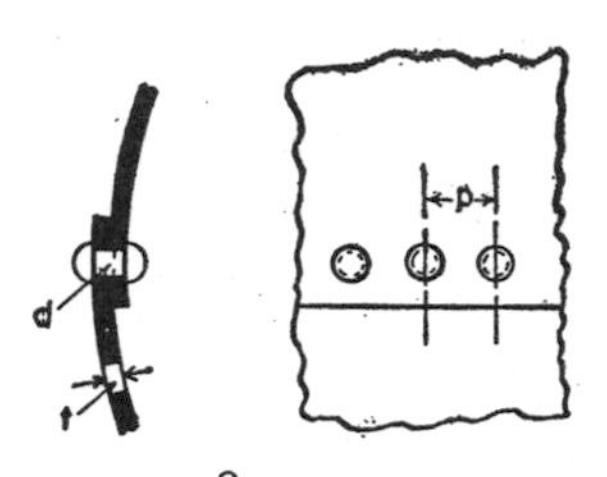

FIG. 238—SINGLE-RIVETED LAP JOINT

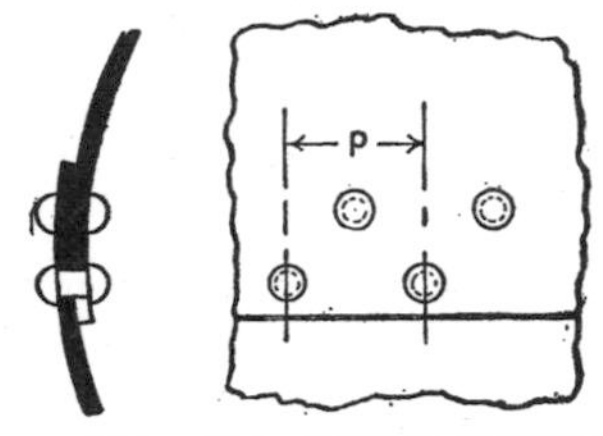

FIG. 239—DOUBLE-RIVETED LAP JOINT

The area to be sheared is the area of a cross-section of the rivet, or

$$\frac{3.1416\,d^2}{4}.$$

The pull which it will take to shear this rivet is the area times the shearing strength:

$$\frac{3.1416\,d^2}{4} \times S_s = \frac{3.1416}{4} \times 45{,}000 = 35{,}343.$$

3. Crushing.—In this case it is common to consider that the area to be crushed is the diameter of the rivet times the thickness, hence

$$dtS_c = 1 \times 7/16 \times 90{,}000 = 39{,}375.$$

The number of pounds it will take to fracture a strip of plate $2\frac{1}{2}$ inches wide and 7/16 inch thick by tension is

$$2\tfrac{1}{2} \times 7/16 \times 55{,}000 = 60{,}155.$$

Hence the ratio of the strength of the joint to the strength of the plate is

$$35,350 \div 60,155 = .588;$$

hence

$$0.588 \times 100 = 58.8 \text{ per cent} = \text{the efficiency.}$$

Now, if the original shell on page 344 is referred to, it will be seen that instead of having a boiler with a factor of safety of 10 it will have only 58.8 per cent of this factor, or approximately 6, which is about the usual factor.

Double-riveted lap joint (Fig. 239).

1. Tearing between two rivets.—The resistance to tearing is

$$(p - d)tS_t = (2\tfrac{1}{2} - 1) \times 7/16 \times 55,000 = 36,095.$$

2. Shearing two rivets.—Instead of shearing one rivet as in the single-riveted lap joint, two are sheared. Hence

$$\frac{2 \times 3.1416\, d^2}{4} \times S_s = \frac{2 \times 3.1416 \times 1 \times 1}{4} \times 45,000$$

is equal to 70,686.

3. Crushing two rivets.—Here again two rivets are considered instead of one, hence

$$2dtS_c = 78,750.$$

The efficiency of this joint would then be

$$100 \times 36,095 \div 60,155 = 60 \text{ per cent.}$$

The same dimensions have been used in this joint as in the previous one for simplicity and comparison. By using a smaller rivet this joint can be made much more efficient.

Test of boilers for strength.—There are two distinct methods of testing boilers for strength. The one which is generally conceded to be best is the hydraulic test; and the other, which is about as safe and sure, and in some cases more so, is the hammer test.

Hydraulic test.—This test consists in filling the boiler full of cold water and then putting pressure upon it to the desired point. This pressure is generally about one and one-half times the working pressure. Since some boilers are designed with a factor of safety of only four or five, if twice the working pressure be put on it there will be danger of rupture to the boiler. With new boilers this test shows all leaks around stays, tubes, joints, etc.; while in old boilers, if they are carefully watched as the pressure increases, it will disclose weakness by bulging in some places and distortion of joints in others.

Hammer test.—The inspector who conducts this test should go over the boiler before it has been cleaned inside and out and carefully note all places where there is corrosion or incrustation. At the same time he should carefully strike all suspicious places a sharp blow with the hammer to detect weaknesses. A good plate will give a clean ring at every blow of the hammer, while a weak one has a duller sound.

Although a boiler may be carefully inspected and tested by both methods, it does not insure it against failures. The greatest strain upon a boiler is due to unequal expansion, and neither of these methods takes this into account.

Some authorities recommend hot water to be used in the test, but there seems to be no advantage in this, since it is the unequal expansion of boilers and not the rise in temperature which causes the failure of certain parts and consequently so much destruction.

FUELS

The fuels most commonly used for making steam are coal, coke, wood, peat, gas, oil, boggasse, and straw. Those used for traction engines and threshing purposes are coal, wood, straw, and occasionally cobs.

Anthracite coal.—Anthracite coal, commonly known as hard coal, consists almost entirely of carbon. It is hard, lustrous, and compact, burns with very little flame, and gives an intense heat. It has the disadvantage when being fired of breaking into small picces and falling through the grates.

Semi-anthracite coal.—This variety has properties that make it to be considered a medium between anthracite and soft coal. It burns very freely with a short flame.

Bituminous or soft coals.—These burn freely and with all gradations of character. Their properties are so varied that they will not permit of classification. Some burn with very little smoke and no coking. This class is generally used in traction engines. Others which coke very freely are good for gas making.

Wood is used only where it is more plentiful than coal. It requires a finer meshed grate than coal and more attention in feeding.

Oil.—In localities where oil is plentiful or where it is cheaper to freight oil than coal, furnaces are fitted for it as a fuel. It has been found that oil burns the best when atomized and mixed with steam. For this purpose a nozzle is constructed so that both steam and oil can flow from it, the steam forming an oily vapor of the oil, which when ignited burns with a very intense heat.

Straw.—In localities where straw is practically worthless and coal and wood are scarce, straw is used as a fuel. It must be handled with care, since too much in the fire box at once is as harmful as not enough.

Value of fuels.—Anthracite and semi-anthracite coals have about the same heating value. Bituminous has a trifle lower value. A cord of hard wood has the same amount of heat in it as a ton of anthracite coal, while a cord of soft wood has only about half that value.

COMBUSTION

The term **combustion** as ordinarily used means the combining of a substance in the shape of fuel with oxygen of the air rapidly enough to generate heat. In all fuels there are hydrogen and carbon, and some mineral matter.

The carbon and hydrogen unite readily with the oxygen of the air, generating heat and light, but the mineral matter remains and forms the ash.

When the carbon of the coal mixes with the oxygen of the air and the mixture is at or above the igniting temperature, combustion takes place and either carbon monoxide (CO) or carbon dioxide (CO_2) is formed, depending upon the amount of air supplied. If the air is insufficient in quantity to furnish enough oxygen to form CO_2, CO will be formed. If the mixture is not hot enough to form complete ignition a great deal of free carbon in the form of smoke is thrown off and is a loss.

Heat of combustion.—Carbon will not unite with oxygen when in the free state until a certain temperature is reached. This temperature is known as the igniting temperature. When the igniting temperature has once been reached and the carbon of the fuel combines with the oxygen of the air, they in turn throw off heat. By experiment it has been found that one pound of carbon burned to carbon monoxide (CO) produces 4,400 B.T.U., and if burned to carbon dioxide (CO_2) 14,650 B.T.U. are produced. One pound of hydrogen united with sufficient oxygen produces 62,100 B.T.U.

Air for combustion.*—By weight, 12 pounds of carbon unite with 16 pounds of oxygen; hence 1 pound of carbon forms

$$28 \div 12 = 2\tfrac{1}{3}$$

pounds CO, or if it be burned to CO_2 it will require twice as much oxygen for each pound of carbon; hence

$$12 + (2 \times 16) \div 12 = 3\tfrac{2}{3}$$

pounds CO_2 for each pound of carbon.

Since in the 3 2/3 pounds CO_2 there is one pound of

*A good discussion of this will be found in Peabody and Miller's "Steam Boilers."

carbon, there must be 2 2/3 pounds of oxygen; hence one pound of carbon requires 2 2/3 pounds of oxygen. As we must have 4 1/2 pounds of air to get one pound of oxygen to burn one pound of carbon to CO_2, it requires pounds of air.

$$2\frac{2}{3} \times 4\frac{1}{2} = 12$$

As there are impurities in all fuels, so that a pound of fuel is not necessarily a pound of pure carbon, there are variations which have to be considered.

Volume of air for combustion.—As before stated, an insufficient amount of air burns the carbon only to CO, while a sufficient amount burns it to CO_2. Instead of having the exact 12 pounds of air for each pound of carbon, as previously computed, it requires an excess for complete combustion. This excess varies from one-half the quantity required for combustion to an equal quantity. Roughly, for each pound of carbon there should be from 18 to 24 pounds of air.

By experiment it has been found that it requires 10 pounds of air for each pound of certain coals, and since 13 cubic feet of air at the temperature it generally enters the fire box weighs 1 pound, for each pound of coal it requires

$$10 \times 13 = 130$$

cubic feet of air without excess. If the excess is 50 per cent, it requires about 200 cubic feet.

Loss from improper amount of air.—If one pound of carbon be burned to CO, there will be 4,400 B.T.U. liberated. If it be burned to CO_2, there will be 14,650 B.T.U. set free. Hence there will be a loss of

$$14{,}650 - 4{,}400 = 10{,}250 \text{ B. T. U.}$$

or

$$100 \times 10{,}250 \div 14{,}650 = 70 \text{ per cent.}$$

This would be a case too rare to be considered and is used only for simplicity. If due caution is practiced in regard to handling drafts, there is very seldom a loss of over 5 to 8 per cent due to lack of air.

On the other hand, if there be too great an excess of air, it would not only furnish oxygen for combustion in sufficient quantities, but the excess would be heated as it passes through the boiler from a temperature of the outside air to a temperature of the flue gases, thus taking up part of the heat which would be transferred to the water. This loss generally amounts to from 4 to 10 per cent.

Smoke prevention.—Black smoke is caused by incomplete combustion. It is generally noticed when starting a fire or when fresh coal is put on. To avoid as much of this as possible, keep the fire hot and feed the coal in small quantities. Do not have the door open longer than is absolutely necessary, as the excess of air cools the fire and instead of burning the CO to CO_2, it passes off as CO or free carbon, which causes the smoke.

HANDLING A BOILER

The flues are made of a soft, tough iron or steel. They are put in place, then expanded with a tube ex-

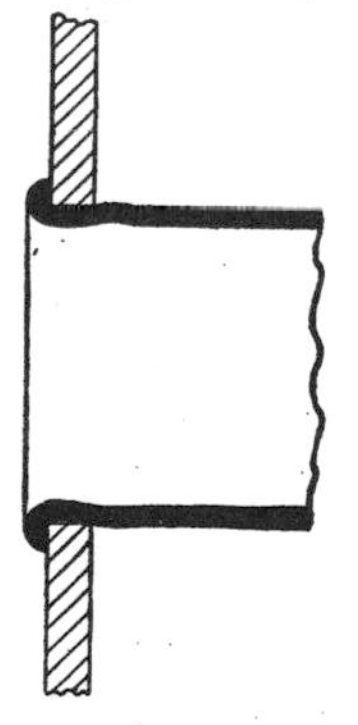

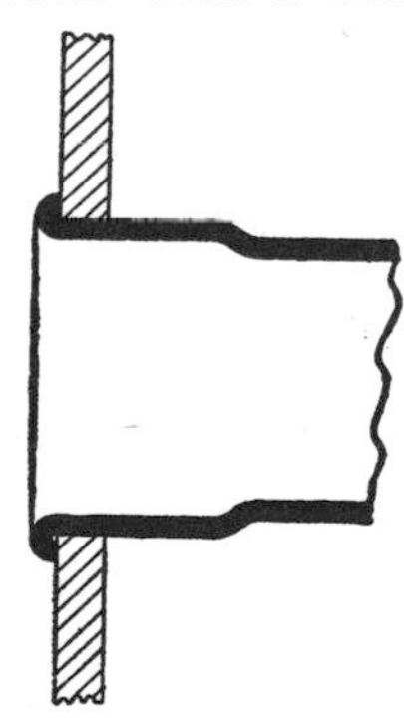

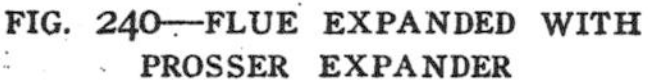

FIG. 240—FLUE EXPANDED WITH PROSSER EXPANDER

FIG. 241—FLUE EXPANDED WITH DUDGEON EXPANDER

pander to a steam-tight joint. The Prosser and Dudgeon expanders are the two types in common use.

The Prosser makes a shoulder on the inside of the sheet as well as on the outside, but permits the tubes to touch only at the outer edges (Fig. 240), while the Dudgeon expander enlarges the end of the tube and causes it to fit the full thickness of the sheet (Fig. 241). Owing to the construction of this type of expander, it is preferable for repair work.

Manholes and handholes.—These are openings in the boiler to permit of cleaning and examining. The use of a manhole is confined to stationary boilers and is generally placed near the top in an opening about 11 × 15 inches. Handholes are generally in the water legs or near the bottom of the boiler. Their accustomed size is about 3 × 5 inches. The plate used to cover these holes is held in place by a bolt passing through a yoke. To secure a tight joint, a ⅛-inch gasket is placed between the plate and the boiler shell of the handholes. The same style of gasket is used for the manholes, but it should be about ¼ inch thick.

Safety valves and steam gauges.—The safety valve should be placed in a pipe by itself, and this pipe should be inspected often for stoppages, etc. The safety valve and steam gauge should be set for the same pressures; that is, if the valve blows off at 110 pounds, the gauge should not read 100 or 120. In case this should happen, do not set the valve to blow off according to the gauge until the gauge has been tested by some gauge known to be correct. During freezing weather the gauge should be taken off every night and put where it will not freeze. Every morning before starting up the safety valve should be tried to see that it neither leaks nor sticks.

Water glass.—There is a cock at each end of the

glass tube. When these cocks are both open the water will pass from the boiler into the glass and stand at the same level as in the boiler, but if either one of the cocks be closed or the pipes leading to the cocks be stopped, the water would rise in the glass and give a false water level. If it is the upper one that is closed, the pressure in the boiler will cause the glass to fill, and if the lower one is closed, the glass will fill with condensed steam. Below this glass is another cock, which is used to drain the glass or blow out the other cocks. By opening this cock when there is pressure and closing the lower one leading to the glass, the upper one will blow out, or if the upper one is closed and the lower opened it will blow out. It is best to try the cocks every morning and see if they are open or free from stoppage. Always have some extra glasses along, for they are likely to break at any time.

Leveling the water column.—Before firing up a boiler a new man should always determine the level of his water in the boiler as compared to the water column. If it is a stationary boiler, take off the manhole cover and fill until the water has reached the lowest limit in the glass. Then continue to fill until the proper height of water has been reached and again note the level in the glass. A good way to mark these points is to file notches in the guard wires which protect the glass.

Should the boiler be traction or portable, it should be set on level ground and leveled up with a level. Then the water column should be leveled the same as in a stationary boiler.

Feed pipe.—There is difference of opinion in regard to the place where the feed pipe should enter the boiler. In horizontal tubular boilers it generally enters near the front end and passes back through the boiler to near the back end before it discharges. In this way the feed water reaches nearly the temperature of the boiler water before it comes in contact with the shell or the tubes. In threshing boilers it generally enters on the side. Sometimes it enters near the bottom through the blow-off pipe.

There should always be a hand valve in the feed pipe near the boiler and a check valve outside of this. The hand valve is placed close to the boiler so that in shutting down in cold weather the water can be shut off. Also if anything happens to the check valve, the hand valve can be closed while the former is being repaired. Where bad water is being used the feed pipe is likely to become choked with scale, and if the pump or injector fails to work it is often well to look in this pipe for the trouble.

Firing.—Before firing up a boiler always see that there is plenty of water. Do not simply look at the glass, but clean the glass and see if it fills immediately. Try the try cocks and see if the water stands the same in them as in the glass. Notice the tubes and grates and see if they are clean.

Firing with soft coal.—Soft coal should not be thrown in in chunks; it should be broken into pieces about the size of a man's fist. Put the coal in quickly and scatter it over the fire as you throw it in. Keep the door open as short a time as possible, so that no more cold air will enter than can be helped. Keep the grates well covered with burning coal so that no cold air will come through them. If the boiler has more grate capacity than needed, do not keep fire on only a part of the grates, but check the fire by closing the drafts. When the fire cannot be kept down in this way without causing incomplete combustion, bricks may be placed over the back end of the grate and to a height equal to the bridge wall.

Some furnaces and fuels require different depths of fire than others. The proper depth can be determined only by trial. Fine coal and a poor draft require a thinner fire than coarse coal and a strong draft. Engineers differ in regard to the best methods for keeping up a fire. Some suggest that it is best to keep the fresh coal near the door, and when it has become coked push it back to the rear, and again throw fresh coal in the front. By this method there is an intense fire maintained at the back of the furnace, and as the partially burned gases pass back they are completely burned. The advantage of this method lies in the fact that complete combustion is secured; consequently there is less smoke, but there is a corresponding disadvantage in keeping the fire door open so long and allowing the furnace to cool slightly.

Cleaning.—Do not clean oftener than necessary. Keep the clinker loosened from the grates between cleaning times. When cleaning large furnaces, rake all the fire to one side and then clean the grates. Rake a part of the live coals back on this side and put on fresh coal. When this is burning well clean the other side in the same manner. To clean small furnaces, crowd the fire back, clean the grates, then rake the fire forward again.

Banking the fire.—Fires are banked to keep the steam from rising when there is a good fire, and also to hold the fire over night. Banking a fire consists in covering the glowing coals with fresh coal or ashes. When banking a fire for the night, crowd the coals to the rear, then fill the front of the furnace with fresh coal, and open the damper over the fire enough to carry off the gases. All drafts should be kept closed. By banking a fire this way it will gradually burn back toward the door, thus keeping the boiler warm, and in the morning there will be a good bed of coals which will start up readily. When a boiler is being used daily, it is considered more economical to bank a fire than to let it go out and then rekindle it in the morning.

Drawing a fire.—Fires are drawn when it is desired to cool the boiler down very quickly or when the water is dangerously low. A fire should never be drawn without first smothering it with ashes, dirt, or fresh coal. Drawing a fire without first doing this causes it to glow up, and for a moment become much hotter than before it was stirred. Never put water in a furnace, as it is liable to crack the grates. It will also produce so much steam that it will either blow back or else blow the fire out the door and make it too hot to work around.

Priming.—When water is carried over from the boiler with the steam the boiler is said to be priming. Priming can always be detected by the click in the engine cylinder, which shows that there is water there. Taking too much steam from the boiler at once, carrying too much water, or not having enough steam space will cause priming. If the cause is too much water, blow out some and then slowly start the engine. Carrying a high steam pressure and keeping the water as low as possible will retard priming to a certain extent.

Foaming is similar to priming, but it is generally caused by dirty or impure water. It can be detected by the rising and falling of the water in the gauge glass and by the engine losing power or speed; also by the clicking in the cylinder. When a boiler foams, the engine should be shut down at once and the water in the boiler allowed to settle. So much water is carried over in the steam that the glass does not show the true level. If after settling down it is found that there is plenty of water over the flues, it will be safe to pump in more, but if the water is low, let the boiler cool down somewhat before filling.

A boiler is more likely to foam with a high-water level than with a low. It is also more likely to foam with low pressure than high. A sudden strain on an engine will sometimes cause the boiler to foam. If a boiler is likely to foam, it is advisable to carry low water and high pressure. Then if it still persists in foaming, shut down and pump in a quantity of water and allow some to run out. This will change the water. If this does not remedy it, the boiler must be cleaned.

Low water.—Should the water happen to get below the danger line in a boiler, immediately cover the fire with ashes, dirt, or even fresh coal, and as soon as it can be drawn without increasing the heat do so. But never draw the fire until it is in this condition. Do not start the feed pump, or start or stop the engine, or open the safety valve. Simply let it cool down. After it has become cool, then examine it for injuries.

If a failure of the injector or pump has caused the water to become low and there is still an inch over the flues or crown sheet, the engine should be shut down and attention given to the feed supply. When the water has become so low as this, do not try to repair the injector or pump with the engine still running, as it will run the water below the crown sheet before it is anticipated and thus make the boiler more dangerous.

Corrosion and incrustation.—It is practically impossible for an engineer to get for his boilers water which does not have some detrimental ingredients. Nearly all hard waters will form some sort of scale. While soft waters do not do this, they do contain acids which act on the boiler and fittings in a harmful manner.

The general impurities to contend with are the carbonates and sulphates of lime. These vary with the location and can be dealt with properly only after experiment. Generally, however, they are thrown down in the boiler in the form of a soft mud and can then be disposed of by blowing out and washing the boiler with a strong stream from a hose. The presence of other impurities, such as oils or organic matter, or even sulphates of lime, makes these lime scales hard and adhesive. Removing the water from the boiler while still hot will cause these scales to bake or dry on the parts, in which case it is very difficult to remove them. Wherever it is possible, run some soft water through the boiler for a few hours before cooling down to clean. The acids will act upon the limes and loosen them from the tubes, etc.

Since the lime impurities of water are thrown down at a temperature of about 200° F., there are devices on the market which allow the feed water to mingle with the exhaust steam. This heats the former to a temperature sufficient to throw out the lime parts.

Boiler cleaning.—It is essential that a boiler be kept clean both inside and out. Authorities have stated that one-tenth inch of scale will require 15 per cent more fuel. Boiler scale is a non-conductor of heat; consequently, the flues must be kept hotter to affect the water as much with scale as without.

The frequency of washing a boiler can only be determined by experience with the water used and the surrounding conditions. Usually a traction boiler should be cleaned once a week, but there are wide variations from this rule.

Often when there is considerable mud in the water it can be blown out by means of the lower blow-off valve. It is good practice to fill the boiler extra full at night; then in the morning when the sediment has settled and there is about 20 pounds of steam, blow off through the lower valve until the proper water level has been reached. When the boiler is in operation the circulation keeps the dirt mixed and it does not avail much to blow off then.

A good way to wash a boiler is to allow it to cool down until one can bear his hand in it; then open the blow-off valve and let the water run out. Remove the manhole and handhole plates and scrape all tubes and the shell with a scraper made for the purpose, then wash well with a hose and force pump.

Cleaning the flues.—Fire tubes should be cleaned at least once a day, and sometimes oftener. This is done by means of a scraper or a steam jet. Scraping should always be done in the morning before firing up. Never do it just after the fire is started, for then the tubes are wet and pasty. If they have to be cleaned while running, do it as quickly as possible and let as little cold air as possible get into them.

Boiler compounds.—Often there are cases where the impurities in boiler waters are such that they form a hard scale. In these cases it is nearly always advisable to use a boïler compound. If the proper compounds are used, they will dissolve the scale and throw it down in the form of a mud. Then it can be blown out. Wherever the scale does not become hard it is very seldom advisable to use a compound.

Wherever a compound is necessary it is best to have a chemist analyze the water and make a compound to suit the case, giving directions as to use and quantity to be used. For traction and small creamery service this is not practical. Soda ash gives very good results for creamery service. It has no offensive odors and is comparatively cheap. Sal soda has also been used with good results. For boilers where steam is used only for engines, kerosene is largely used. Kerosene is also good to remove scale already formed. Where a sight-feed lubricator is available, kerosene may be fed through it, but when not the kerosene may be put into a boiler before filling. The kerosene floats, and as the water rises it adheres to the sides and tubes. Avoid using a compound except when absolutely necessary.

Blister.—A blister in a boiler is identical with a blister on the hand. On account of imperfect material or dirt, the metal will separate and one part will swell. Wherever there is a blister it is best to cut this part out and patch. If the blister is around the fire, a new half sheet should be put in.

Bag in a boiler.—A boiler is likely to bag if dirty, or if a quantity of oil has found its way into it. The oil will stick in one place and keep the water away. Then the fire will overheat this place and the inside pressure force it out. In forcing out the place it breaks the oil scales and allows the water to run in and cool it off. Sometimes it is best to put in a new half sheet where a bag is formed, but often it can be repaired by heating the place and driving it back.

Cracks sometimes form in the flue sheet because the flues are expanded too much. They are often formed in riveting. Whenever a crack is discovered it can be mended by drilling a hole in the end of the crack and putting in a rivet. This keeps the crack from getting larger; then the crack can be filled in.

Laying up a boiler.—In laying up a boiler, always clean it thoroughly. Scrape and wash it inside and out, and then paint the outside with black asphaltum or graphite and oil.

CHAPTER IV

STEAM ENGINES

Early forms.—Hero of Alexandria is given credit for being the first man to use steam as an agent to convert heat energy into mechanical energy. He produced an æopile which operated with steam upon the same principle that our present-day centrifugal lawn sprinklers work with water.

History gives us ideas which were advanced by certain men, but nothing of importance after Hero's machine until 1675, when, conjointly, Newcomen, Calley, and Savery invented what has been known as the Newcomen engine. Fig. 242 is a drawing of this engine as it was used for pumping water. *A* is the pump plunger and is always held down by the weights *B*. The steam, after being generated in the boiler *C*, is passed through valve *D* to the cylinder *F*. The piston *H*, which is up as the steam enters, is connected with the pump by means of the walking beam *I*. When the cylinder *F* is filled with steam, the valve *D* is closed and the valve *E* opened, letting in a jet of water from the previously filled tank *G*. As the water enters the cylinder it condenses the steam *F*, thus producing a vacuum in the cylinder, consequently the atmosphere will act upon the piston *H* and force it down. As it forces the steam piston down it raises the piston *A*, and with it the water.

After Newcomen, Watt produced probably the most important improvement of the steam engine. It was in 1769 that he got out an engine which would not condense the steam in the working cylinder, and by so doing cool off the walls, but he condensed it in separate vessels, which produced a continuous vacuum. The same principle as that of Watt is in use in the condensing steam engine of to-day, the only changes being in the mechanism for admitting and releasing the steam, in mechanical make-up and methods whereby labor in the machine shop is reduced.

The present engine.—The working parts of the present engine are all of the same general plan, with different designs for carrying out the actions. The principle is that of a cylinder separated into two parts by a piston. There is a valve connected with the cylinder by

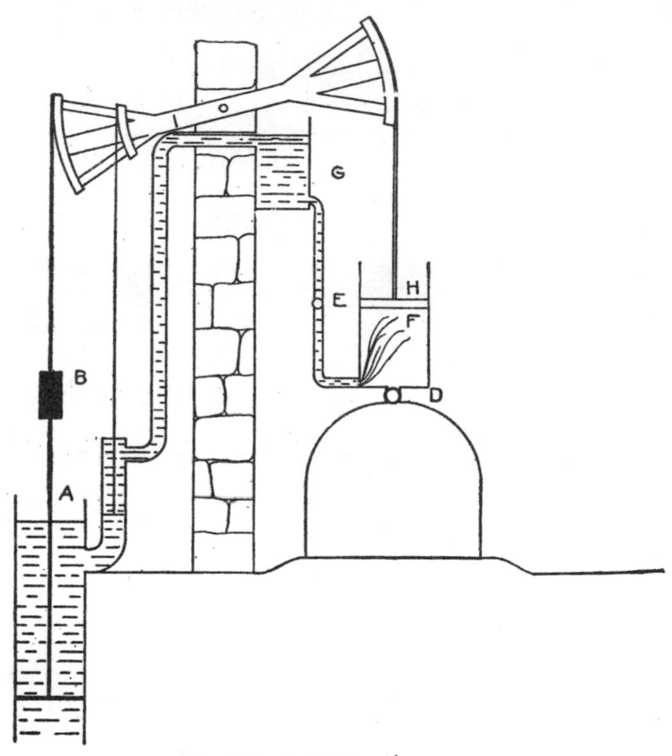

FIG. 242—NEWCOMEN'S ENGINE

means of which the steam is thrown from one side to the other. This valve also conducts the exhaust steam out of the cylinder. In Fig. 243, *A* is a steam chamber which receives the steam from the boiler. *B* is the valve which slides back and forth on the valve seat *J*. The valve *B*, situated as it is in this figure, allows the steam to pass from the steam chest *A*, through the steam port *C*, into the front end of the cylinder *D*, and press against the piston *E*. This forces the piston through the cylinder toward the end *F*. At the same time the steam which

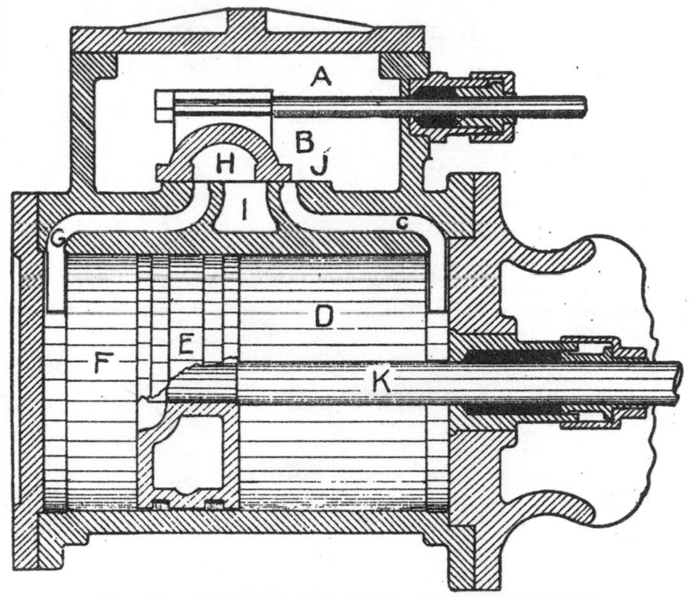

FIG. 243—CYLINDER AND VALVE OF STEAM ENGINE

has been previously admitted to the end of the cylinder *F* is forced out through the cylinder port *G* into the exhaust chamber *H*, and out through the exhaust port *I* into the air. By the time the piston *E* has reached the end of the stroke the valve *B* has reversed its position so that the steam chest *A* is connected with the end of the cylinder *F* by way of the steam port *G*. The exhaust port *I* is now connected with the exhaust end of the cylinder *C*, hence as the steam enters the cylinder at the end *F* it drives the piston toward the end *D*.

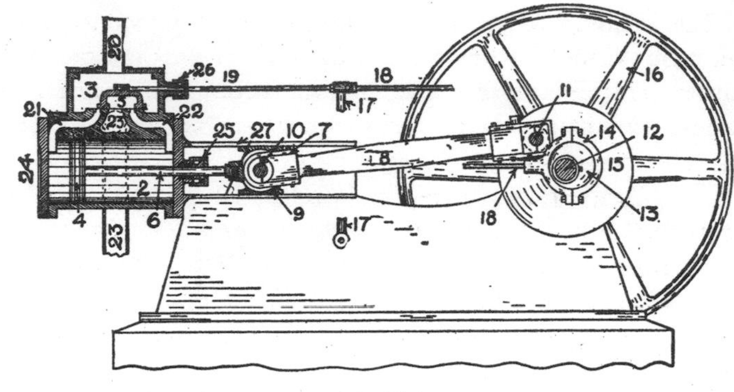

FIG. 244

1. Base.
2. Cylinder.
3. Steam chest.
4. Piston.
5. Valve.
6. Piston rod.
7. Crosshead.
8. Connecting rod.
9. Crosshead shoe.
10. Wrist pin.
11. Crank pin.
12. Crank shaft.
13. Eccentric.
14. Eccentric strap.
15. Crank disk.
16. Flywheel.
17. Valve rod guide.
18. Eccentric rod.
19. Valve rod.
20. Steam inlet pipe.
21–22. Steam ports.
23. Exhaust pipe.
24. Cylinder head.
25–26. Packing boxes.
27. Guides.

Classification of steam engines.—

- Speed
  - High
  - Low
- Disposition of Steam
  - Condensing
  - Non-Condensing
- Number of Expansions
  - Simple
    - Single
    - Double
  - Compound
    - Tandem
    - Cross
    - Twin
- Speed Regulation
  - Throttling Governor
  - Automatic
  - Corliss
- Kind of Work
  - Stationary
  - Marine
  - Locomotive
    - Rail
    - Traction
- Pressure on Piston
  - Single Acting
  - Double Acting

The classes of engines generally used in agricultural pursuits would be known as high-speed, non-condensing, either simple, single or double, or compound tandem or cross, throttling governed, either stationary or locomotive traction and double-acting.

Generation of steam.—Enclose 1 pound of water at a temperature of 32° F. in a cylinder under a movable frictionless piston. Suppose the piston to have an area of 1 square foot, but no weight other than the atmospheric pressure. Apply heat to the water and the following results will be noted:

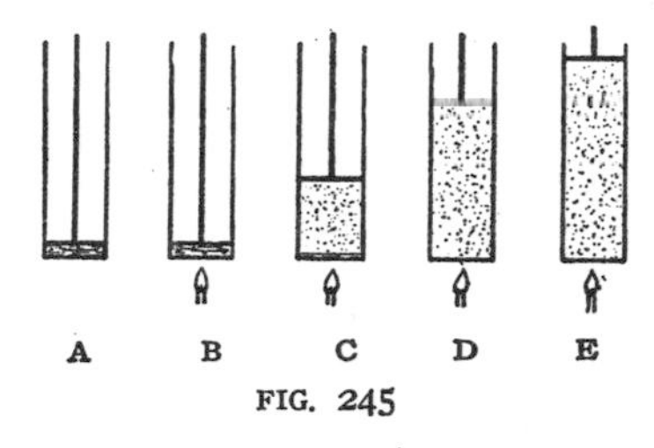

FIG. 245

A, one pound of water at 62° F.; *B*, one pound of water at 212° F., but lacks heat enough to turn it into steam; *C*, is saturated steam in contact with the water; *D*, one pound of steam at 212° F.; *E*, one pound of superheated steam.

1. The temperature rises, but the piston remains in the same position until a certain temperature is reached. When the piston commences to rise the degree of temperature is known as the boiling point. This point varies with the pressure. If the pressure bearing on the piston had been 10 pounds to the square inch instead of 14.7, the boiling point would have been reached at a lower temperature, and if the pressure had been 20 pounds to the square inch, the boiling point would have had a higher temperature.

2. As soon as the water has reached the boiling point, though heat still be applied, there is no further rise in

temperature, but steam forms and the piston gradually rises. Since the water is passing into steam, it must be disappearing. During formation the steam and the water remain at the same temperature as the water was when steam commenced to form. The heat which has been continually added has been used to convert the water into steam and is known as **latent heat.**

3. After all the water has been evaporated, if heat be still applied the temperature of the steam will commence to rise and the piston will also continue to rise. Since the steam is not now in contact with the water and is hotter than the steam was when formed and in contact with the water, we have **superheated steam**; in other words, steam which is heated above the temperature of the boiling point of water, which corresponds to the pressure at which it is generated.

Saturated steam is steam at its greatest possible density for its pressure. It is invisible and must contain no water in suspension; in other words, it must be dry and still not be superheated. The temperature of saturated steam in the presence of water is the same as that of the water, and for steam of a given temperature there is only one pressure. If the temperature increases and the volume remains constant, the pressure does likewise, for as the temperature increases more water is evaporated, or if the temperature decreases the pressure does also and some of the water is condensed.

Total heat of steam is made up of two components, heat of the liquid and latent heat.

Heat of the liquid is the amount of heat there is in water at the temperature of the steam.

Latent heat is the amount of heat required to evaporate 1 pound of water at a given temperature into steam at the same temperature. It is made up of two components.

One is the heat required to overcome the molecular resistance of water to changing from the liquid state to the gaseous. This is known as internal latent heat. The other component is the heat required to overcome the external resistance or pressure.

Volume and weight of steam.—The weight of a cubic foot of steam at 212° F. is 0.03758. If the temperature be increased to 337°, which corresponds to a gauge pressure of 100 pounds, the weight of a cubic foot will be 0.2589 pounds. By increasing the weight of steam we decrease the volume; i.e., the volume of 1 pound of steam at 212° is 26.64 cubic feet, but at 337° it is only 3.86 cubic feet. Hence when it is stated that steam has a volume of so many times the volume of an equal weight of water the temperature or pressure of the steam must be known. Often in testing a steam boiler it is assumed that as many pounds of steam are evaporated as there have been pounds of water fed to the boiler. This is an erroneous assumption, for there is always a certain per cent of the steam which is not steam but water in suspension. This, of course, will make the boiler appear to be generating more steam than it really is, but when this wet steam comes to the engine it will be charged against the engine as using all steam and consequently much more than is necessary, when in fact it is not using so much steam as is recorded, but is passing water through the cylinder.

Expansion of steam.—When saturated steam is used in an engine without expansion only about 8 per cent of the heat expended is converted into useful work. By not admitting steam into the cylinder for the full length of stroke, as shown in a previous part of this chapter, but by cutting it off during the first part of the stroke and allowing it to expand during the remaining

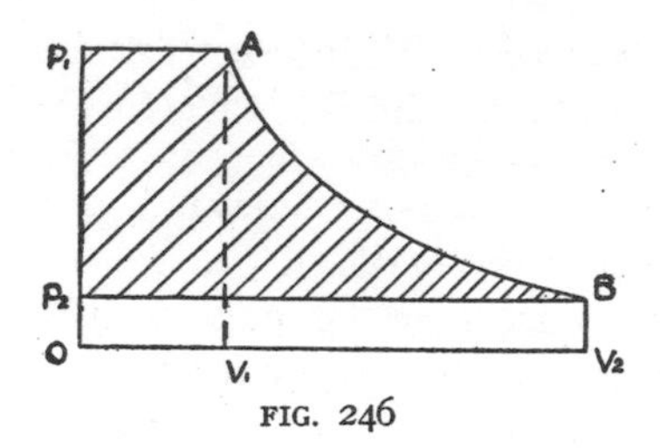

FIG. 246

part of the stroke, more work can be obtained from the same amount of steam.

In Fig. 246 let the distance OV_2 represent the length of stroke, OP_1 the pressure of steam as it enters the cylinder and while in communication with the boiler. If the piston starts at the point O and travels to V_1 with the valve wide open, steam will continue in the cylinder at the pressure of the boiler, i.e., the pressure at A will be the same as at P_1 and the line P_1A will be parallel to the line OV_1. Now, if steam is cut off at V_1 and no more allowed to enter, the pressure will fall as fast as the steam expands and the line AB is formed. During this part of the stroke all the work which is done in the cylinder is due to the expansion of the steam which was admitted during the first part of the stroke. When the piston reaches V_2 the steam is exhausted against a back pressure of OP_2.

The work done during the admission of steam is represented by the area OP_1AV_1, and is all the work this amount of steam would do if it had not been allowed to expand.

The work done during expansion is represented by the area V_1ABV_2.

The total work done by the steam is the sum of these two areas, or OP_1ABV_2.

Then, of the total work done by the steam that represented by the area V_1ABV_2 is gained by using the steam expansively.

Losses in a steam engine cylinder.—Only 2 to 10 per cent of the total heat supplied to a non-condensing steam engine goes into useful work. In multiple-expanding steam engines this percentage is often raised as high as 20. The rest of the energy is lost by radiation, condensation in the cylinder, and the amount carried away to exhaust. The temperature of the walls of the cylinder rises and falls as live steam enters and expands to the pressure of exhaust; in other words, the cylinder walls have practically the same temperature as the exhaust steam, so when the live steam enters it heats the walls to a temperature nearly equal its own. This then is the loss due to radiation. As the steam expands in the cylinder there is a great deal of it which condenses. Due to this condensation, the latent heat of the steam is thrown off, doing no work. Not only is all the heat left in that part of the steam which entered the cylinder to fill the piston displacement lost when release takes place, but about one-third of the steam which enters the clearance space is a total loss. Hence the smaller the clearance volume the more economical the engine.

Slide valve.—The slide valve is the most common method for regulating the admission of steam to and exhaust of the steam from a steam engine cylinder. Its functions are: (1) admission of the steam to the cylinder to give the piston an impulse; (2) to cut off the supply of steam at the proper point; (3) to open a passage for the escape or exhaust of the steam from the cylinder; (4) to close the exhaust port at the proper time to retain enough steam in the cylinder to give the piston a cushion.

Lap of valve.—When the valve is in mid position (Fig. 247) the amount it laps over the edges of the steam port is known as lap. The amount which the valve laps over the outside is outside lap, and that which it laps over the inside is inside lap.

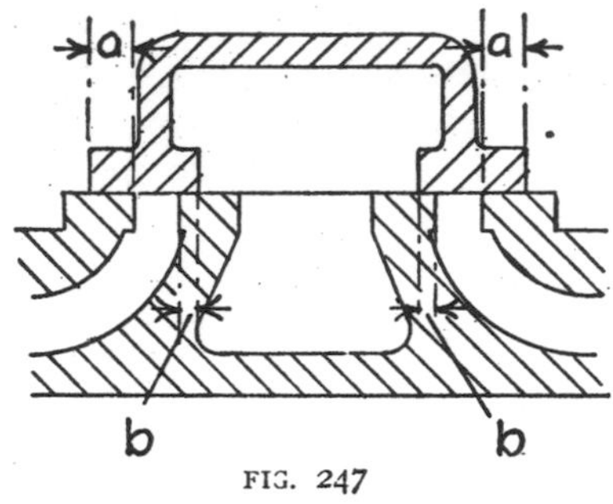

FIG. 247

Object of lap.—Lap is put on the slide valve to secure the benefit of working steam expansively. If a valve has no lap, steam will be admitted the full length of the stroke and allowed to escape to the exhaust at boiler pressure. By the application of lap, steam is cut off from the boiler when the piston has traversed from three-eighths to five-eighths of the stroke, and as the piston completes the stroke the steam does work by expanding.

Lead is given to a valve to admit steam to the cylinder just before the piston reaches the end of the stroke. By so doing a cushion is produced in the cylinder upon which the piston acts and this saves a jar. Lead not only produces this cushion effect, but also causes the port to be partly opened so that a full amount of steam can be admitted to the cylinder the instant the piston starts on its return stroke. Lead affects the ex-

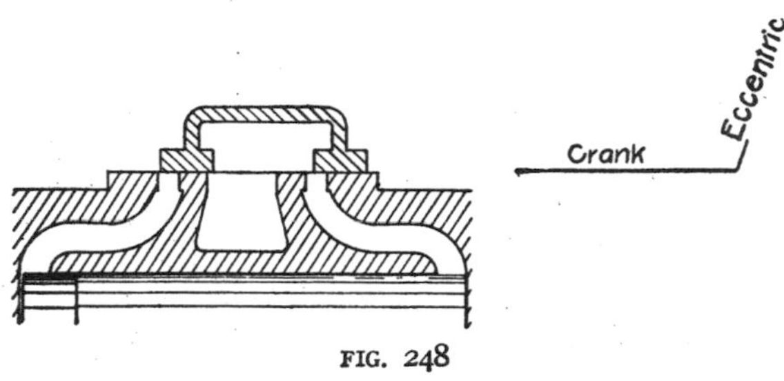

FIG. 248

haust port by having it open in time for the exhaust steam to be sufficiently released so that at the instant the piston starts on the return stroke there is no back pressure. Fig. 248 shows the lead, both inlet and ex-

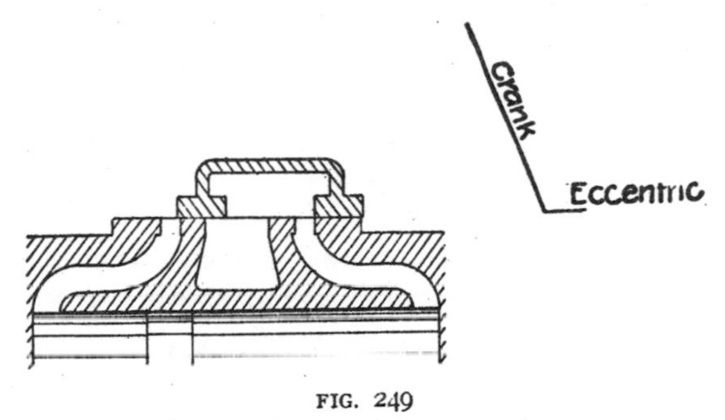

FIG. 249

haust. Fig. 249 shows the valve when at its end of the stroke, showing that the exhaust port is completely opened, but that the inlet is not necessarily so. Fig. 250 represents the position of the valve when the piston is at the opposite end of the stroke. It will be noticed that

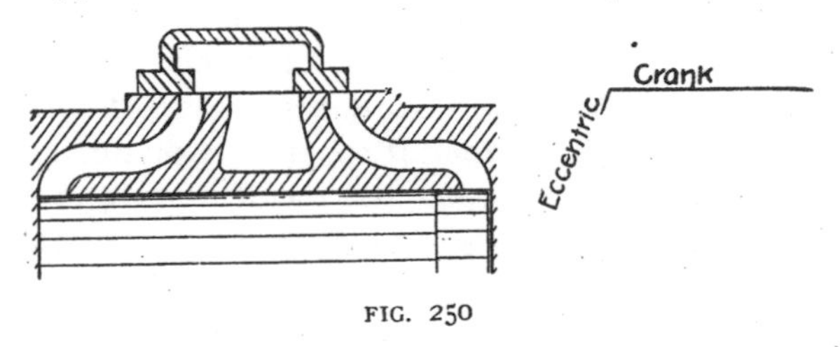

FIG. 250

the lead in this case is the same as that in Fig. 248. This should be true in all engines.

lead of 1/32 inch is about proper for most engines. Too much lead in a valve allows steam to enter the cylinder so soon that the piston has to complete its stroke against boiler pressure, hence a loss of energy. Also where there is too much lead the exhaust port is likely to open so soon that the steam is released before it expands as much as possible. Again, if it has not sufficient lead there will be no cushioning effect, and in addition sufficient steam will not have entered the cylinder by the time the piston starts on the return stroke to produce the maximum pressure.

Eccentric.—The eccentric is a mechanism often used where it is impossible to use a crank. The eccentric of a steam engine consists of a disk or sheave fastened to the crank shaft in such a manner that it is eccentric or out of center with the center of the shaft. Around this sheave is the eccentric strap, which is so adjusted that there is a free and smooth bearing surface between the two. The eccentric rod, which actuates the valve, is attached to a strap and gives to the valve a reciprocating motion similar to that of the piston, but on a reduced scale. The throw of the eccentric, which is also the travel of the valve, is twice the distance from the center of the eccentric to the center of the shaft. In other words, it is the same as that of a crank whose length of arm is equal to the eccentricity of the eccentric.

Angle of advance.—On a slide-valve engine, with the valves properly set when the engine is on dead center, the center line of the eccentric will not be at right angles to the crank, but will be at an angle greater than a right angle. The difference between this angle and a right angle is known as the angle of advance.

The size of this angle varies with different engines, but it is generally from 10° to 20°. The object of the angle of advance is to give the engine lead, and to vary the lead means to change the position of the eccentric on the shaft. Changing the position of the eccentric changes the angle of advance.

In Fig. 251 let AB be the travel of the valve, OA the position of the crank, and OC the position of the eccentric. Then the angle COD, or θ, is the angle of advance. A perpendicular let fall from C to OB gives the distance OE, which designates the position of the valve. In this instance it also gives the lead, i.e., OE, is the lead of the valve. If the position of the crank is changed from OA

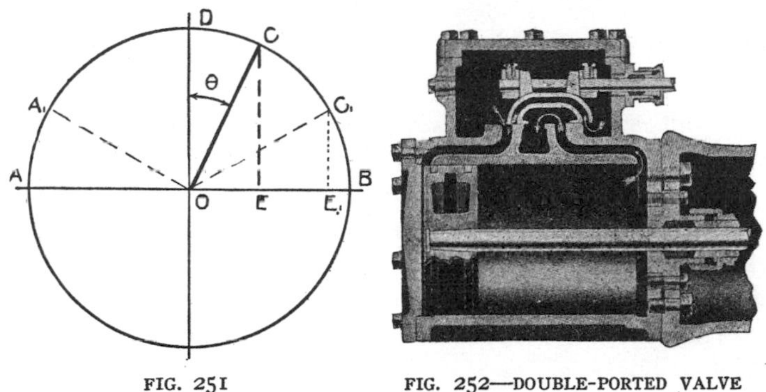

FIG. 251

FIG. 252—DOUBLE-PORTED VALVE

to OA_1 the valve will move the distance OE_1 or a total distance of $EE_1 + OE = OE_1$.

Double-ported valves.—The common slide valve has to travel so far in opening a steam port that there is considerable wire drawing of the steam as it enters the cylinder; also it does not permit a free release of the exhaust steam. Some manufacturers are putting in their engines a double-ported valve (Fig. 252) which gives about the same port opening as the simple slide valve and with only half the travel.

Balanced valve.—By inspecting Fig. 243 it will be noticed that there is high-pressure steam all over the outside of the valve and none on the inside. This excessive pressure on the outside causes a large amount of friction between the valve and the valve seat. To overcome this excessive friction balanced valves are now made. Some have on the back a friction ring, which is held against the steam chest by coil springs or live steam in such a manner that the steam does not get behind the valve. Other valves are so constructed that the high pressure steam is kept from the back of the valve by means of pieces of strap steel working in grooves in the back of the valve. These pieces of steel are generally held out against the steam chest cover by means of coil springs. Fig. 253 illustrates this type of valve.

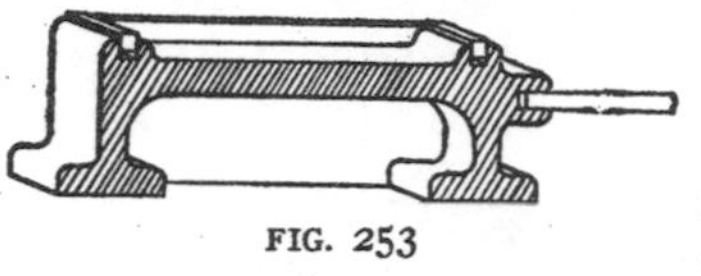

FIG. 253

Piston valve.—The piston valve is probably the most effectually balanced valve. The principle of this valve is the same as that of the common slide valve, but instead of having a seat it is cylindrical in form and has packing rings the same as a piston, making it steam tight (Fig. 254).

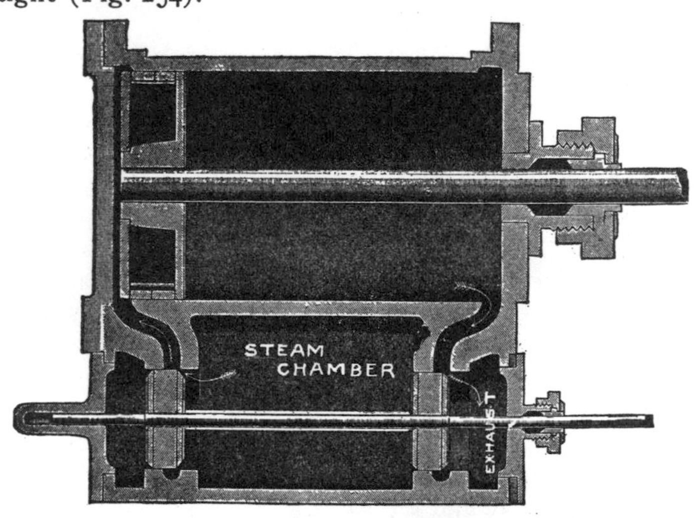

FIG. 254—PISTON VALVE

Dead center.—An engine is on dead center when a straight line passing through the centers of the crosshead and crank shaft will pass through the crank pin. If an engine is on dead center it will not start, although the ports may be open. Locomotives and often traction engines have two cylinders with their cranks at right angles, so that one or the other will always be off center, and consequently will start without turning the wheel by hand.

Locating dead center.—When the crank is passing dead center the piston moves so slowly that a movement of 2 or 3 inches of the crank is hardly perceptible on the piston. This, however, is not true of the valve, for when the crank is passing dead center the valve is moving its fastest, consequently it is essential that dead center be definitely determined. About the simplest and most accurate method for putting an engine on dead center is by means of a tram (Fig. 255). At some convenient place in the engine frame make a clear, sharp-cut center-punch mark, and with the flywheel about one-eighth

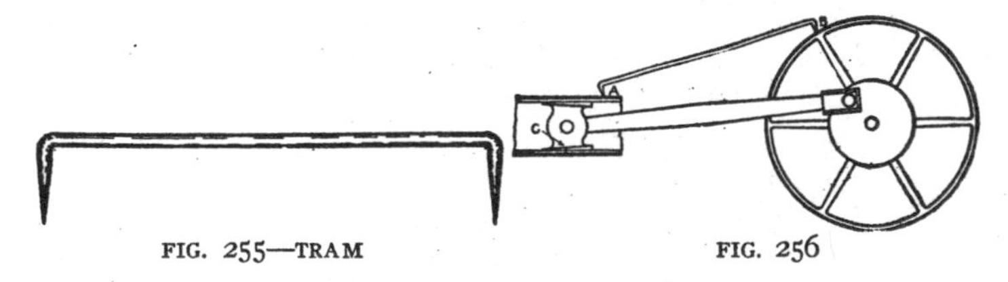

FIG. 255—TRAM FIG. 256

revolution off center make another center-punch mark in the wheel. Set the tram in the center-punch marks as shown in Fig. 256. Now with a sharp knife make a mark *C* across the intersection of the crosshead an dthe guide. Turn the wheel down until the mark on the crosshead and the guide come together again, then make another mark in the wheel so that the tram will drop into it as in Fig. 257. Having done this, find the point *E*, midway between the two marks on the flywheel, and make a punch mark there. Turn the wheel until the tram drops into this mark (Fig. 258), and the engine will be on dead center. To find the opposite dead center do likewise or

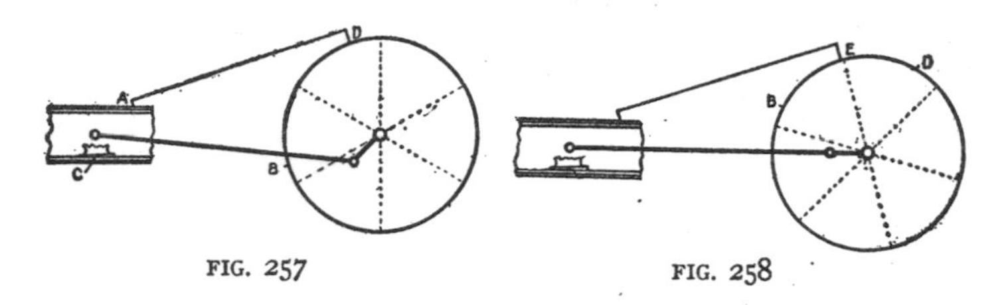

FIG. 257 FIG. 258

measure half around the wheel. When it is inconvenient to measure on the flywheel, the crank disk can often be used.

Setting the slide valve.—To set the slide valve, remove the steam chest cover and put the engine on dead center. Turn the eccentric on the shaft until it is 90°, or a quarter of a revolution, ahead of the crank in the direction the engine is to run. Now adjust the valve on the rod until it is at its center of travel, then again move the eccentric ahead, this time only until sufficient lead is obtained. Fasten the eccentric to the shaft and tighten up the lock nuts on the valve; then turn the engine over to the other dead center and see if both sides have the same lead. If the lead is the same in both ends, the valve may be set. If there is more lead in one end than the other, move the valve on the rod an amount equal to one-half the difference. If now the valve has too much or too little lead, the eccentric should be slipped forward or backward, as the case may require.

Moving the eccentric in the shaft increases or diminishes the lead, depending upon the direction it is moved.

Moving the valve on the rod increases or diminishes the difference in lead.

If an engine has a rocker arm pivoted in the center, move the eccentric in the opposite direction. Otherwise proceed in the same manner as without the rocker arm.

Reversing a simple slide valve engine.—To set the valve of a simple engine so that the engine will run backward, or, as is often termed, **under,** remove the steam chest cover, set the engine on dead center, and ascertain the lead. Now loosen the eccentric from the shaft and turn it backward until the lead is again the same as before. The distance which the eccentric is to be turned backward should be 180° plus twice the angle of advance (Fig. 259).

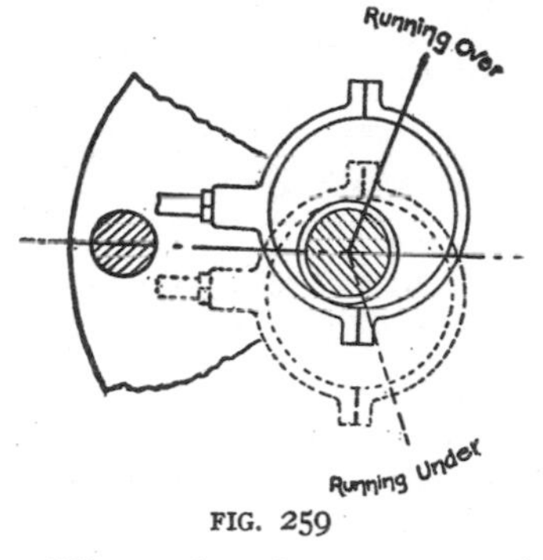

FIG. 259

The valve does not need to be moved on the rod, nor the rod lengthened or shortened. The only caution necessary is to be sure that the lead is always on the end the piston is on when the engine is on center.

An engine running backward or under will do just as much work as one running forward or over, but when it is running over the pressure of the crosshead is always down, while when it is running under the weight of the crosshead and connecting rod is down, but the pressure caused by the steam on the piston and the angle of the connecting rod and piston rods will be up; hence there are two forces working in opposition at the crosshead, and this will cause an up-and-down pound. Not only this, but if an engine runs over, this force will all be exerted upon the engine bed and not the frame.

Reversing gears.—Since the simple engine cannot be reversed without stopping and using time, engines which have to be reversed often and quickly are provided with reversing gears. That is, they are arranged so they can be reversed with a lever. There are two general classes of reversing gears, the double-eccentric and the single-eccentric.

Hooking up an engine.—Some engine makers designate their reversing gears as expansion gears. Such gears are simply reversing gears which can be used so that the steam works on expansion. Reversing gears are actuated by means of a lever which works in a quadrant. When the lever is in one half of the quadrant steam is admitted so that the engine runs under, and when in the other half the engine runs over. These gears are generally so constructed that if the engineer wishes his flywheel to run in a direction away from him he moves the lever in the direction the wheel turns, and if he wishes the wheel to run toward him, he moves his lever in that direction. Some engines are connected up in the opposite manner. When an engine is carrying an overload, the lever is thrown into the last notch in the quadrant and the piston receives steam nearly the full length of the stroke. Although this has to be resorted to in some instances, it is not an economical way to run an engine, as the steam has no chance to expand. When an engine is running on full load, that is, when it is doing only its rated capacity of work, the lever should not be in the end notch of the quadrant, but should be somewhere between the end notch and the middle. By having an engine hooked up, steam is cut off in an earlier part of the stroke and consequently works on expansion the remaining part.

Double-eccentric reverse or link-motion reverse.—

There are several types of this reverse, but probably the Stephenson link is the most popular. It will be described here. In Fig. 260, *A* is the quadrant over which the reverse lever *B* works. The reverse lever *B*, acting through the rocker arm *C*, raises and lowers the link *H*. *F* and *G* are eccentric rods connected at one end with the eccentrics *D* and *E*, respectively, and at the other end with the ends of the link *H*. *I* is a block which is attached to this end of the valve rod and is worked over by the link *H*. With the reverse lever in the position in which it now is, the eccentric *D*, through the rod *F* and block *I*, actuates the valve. By throwing the reverse lever to the other end of the quadrant, the link is raised so the eccentric *E*, through the rod *G* and the block *I*, actuates the valve. It will be noticed that the angles of advance of these two eccentrics are practically the same as they were for the two positions of the eccentric in Fig. 259, where the simple engine was reversed. Thus it is seen that the engine has been reversed by simply shifting the motion of the valve from an eccentric which runs the engine under by means of the link *H* and the block *I*, to an eccentric which runs it over. If the reverse lever is hooked up in the middle notch of the quadrant, the block *I* will be acted upon by both eccentrics, one acting in one direction and the other oppositely; consequently there is only a very slight movement of the valve.

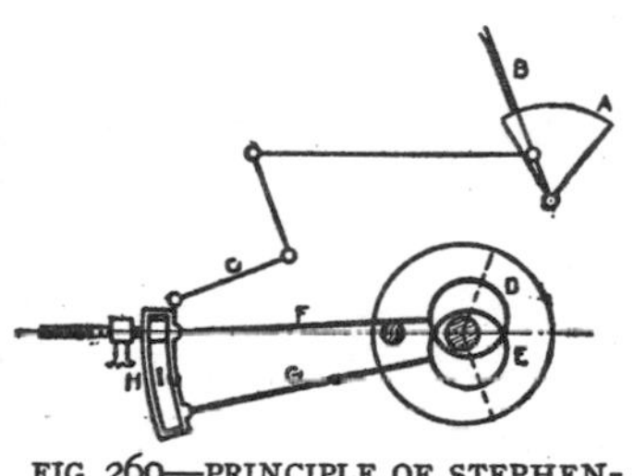

FIG. 260—PRINCIPLE OF STEPHENSON LINK MOTION

Setting the double-eccentric valve.—Put the engine on dead center and drop the link down as far as possible and still have clearance between the link and the block; then set the valve in the same manner as a simple slide valve. To set the other eccentric, raise the link and proceed in the same manner, but remember the engine is to run in the opposite direction.

Single-eccentric reverse gear.—Like the double-eccentric reverse, there are several types of the single-eccentric reverse, but the Woolf reverse gear, being the

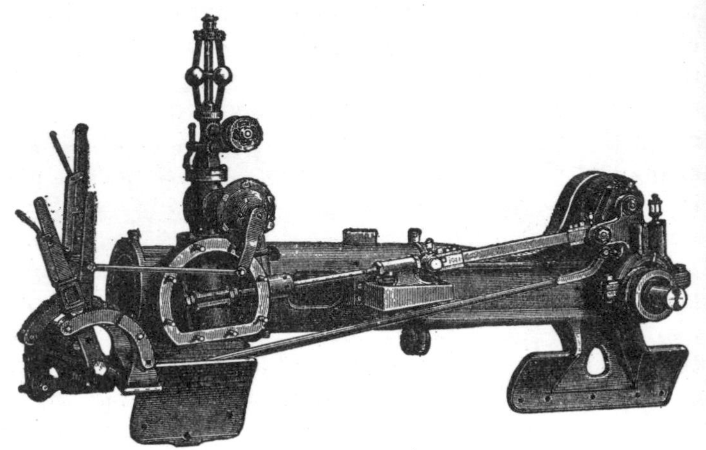

FIG. 261—WOOLF REVERSE GEAR

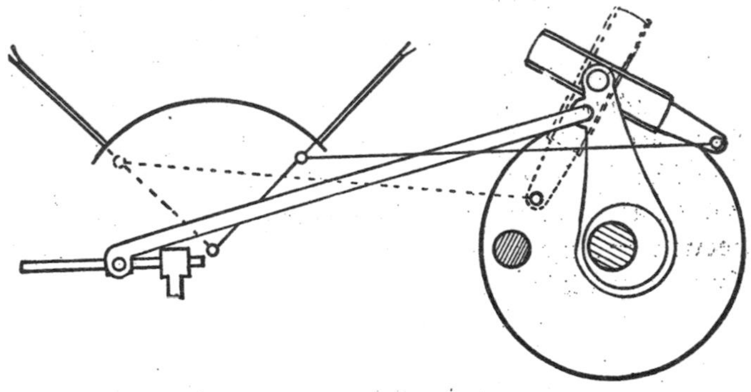

FIG. 262—PRINCIPLE OF WOOLF REVERSE GEAR

most common, will be discussed here. This reverse gear (Fig. 261) has few parts to wear and get out of order and may be set so that steam can be used on expansion.

It will be noticed that in this reverse gear (Fig. 262) the throw of the eccentric is set opposite to the crank instead of about at right angles to it, as shown with other gears. By moving the lever from one end of the quadrant to the other the guide *A*, which carries the roller *B*, changes position as shown by dotted lines. This causes the valve to move in the opposite direction. All types of reversing gears have some mechanical means of operating the throw of the valve. This is equivalent to changing the position of the eccentric in the shaft, and if one method of setting the valve is mastered all others will be easily picked up.

Angularity of connecting rod.—Due to the angularity of the connecting rod, the piston of an engine travels faster and farther while the crank is passing through the half of its rotation nearer the cylinder than it does while the crank travels the opposite half of its rotation. By reference to Fig. 263 it will be noticed that the crank has traveled only half its distance and the piston has passed over more than half its stroke. As the crank passes through the other half of its revolution, which it does in the same time as it did the first half, the piston travels as much less than half its stroke as it traveled more than half during the first revolution of the crank, consequently does not travel nearly as fast during this half of the time as it does during the other half. Because of this unequal travel of the piston one end of

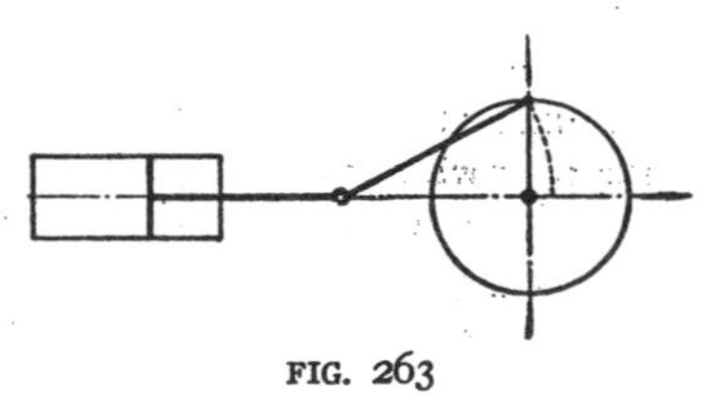

FIG. 263

the cylinder is doing more work than the other, and as a result there is excessive vibration and unequal strain in the parts. It is impossible to change the connecting rod, but there are now valve gears on the market which partly rectify the defect by the manner in which they admit the steam. Owing to mechanical complications which arise, it is still a question as to the advisability of putting these valves on small engines.

The indicator diagram.—Fig. 264 is an ideal indicator diagram and can be described as follows: The line *xy* is traced on the paper with no pressure in the cylinder, i.e., it is the atmospheric line.

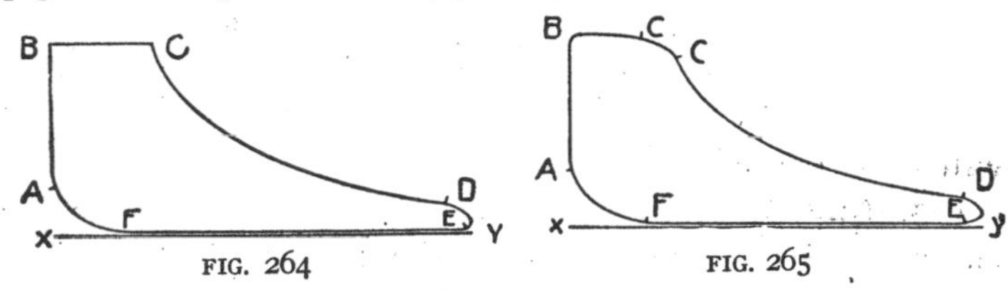

FIG. 264 FIG. 265

The point *A* shows when steam commences to enter the cylinder. Point *B* is the maximum pressure and the time when the steam port is opened its full amount. From *B* to *C* the port is open, and the pressure is the same as *B*. At *C* the cut-off takes place and the steam works on expansion. At *D* the exhaust port opens, and from *D* to *E* the pressure drops to the pressure at which the steam exhausts to the air. From *E* to *F* is back pressure, due to exhaust. At *F* compression takes place and lasts until *A* is reached.

The different parts of the diagram are known as follows:

xy. Atmospheric line,
AB. Admission line,
BC. Steam line,
CD. Expansion line,
DE. Exhaust line,
EF. Back pressure line,
FA. Compression line,
A. Point of admission,
C. Point of cut-off,
D. Point of release,
F. Point of compression.

There are mechanical difficulties which must be taken into consideration; hence the diagram as usually obtained from a steam engine cylinder is not like Fig. 264, but is like Fig. 265. Here the corners are rounded off, due to wire drawing and slow-acting valves. The line *BC* drops, due to the resistance of steam moving through the boilers. The point *C* is not a sharp one, since the valve cannot move quickly enough to cut off steam instantaneously, but commences to cut off at *C′*, and complete cut-off takes place at *C*. This fall in pressure after the valve commences to cut off and before it completely cuts off is known as wire drawing. Often the exhaust valve does not open soon enough for the pressure to fall to the back pressure line before the piston starts in the return stroke; hence the line *DE* of Fig. 264 is more like the line *DE* of Fig. 265.

Attaching indicator to engine.—Where indicator diagrams are to be taken from engines of 100 H.P. or more it is better to have two indicators, one for each end of the cylinder; but for engines of a capacity such as are used on the farm or in creameries one indicator connected to both ends of the cylinder by means of a three-way cock is fully as accurate as two. If there are no holes for attaching the indicator when the engine comes from the factory, drill into each clearance space *AA* (Fig. 266) of the cylin-

FIG. 266—ATTACHING AN INDICATOR TO AN ENGINE

der a hole of sufficient size to thread for 3/8-inch or 1/2-inch pipe, and by means of pipe fittings connect up to the three-way cock *B*. The connection on the indicator will screw into the cock at *C*. Since the throw of the indicator drum is only about 3½ inches and the stroke of the piston is 8 to 20 inches, the length of stroke of the piston has to be reduced to that of the indicator. There are several mechanisms for this purpose, some of which come with the indicator (Fig. 267). If a reducing motion has to be devised, probably that shown in Fig. 268 is the most simple.

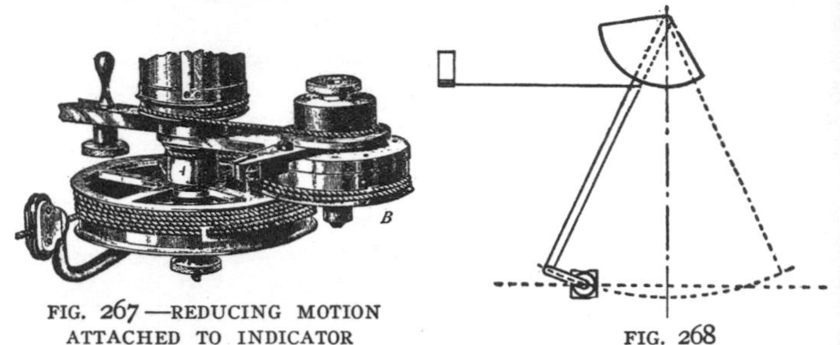

FIG. 267—REDUCING MOTION ATTACHED TO INDICATOR

FIG. 268

Taking indicator diagrams.—To take an indicator diagram the string after being hooked up should be of proper length to give the indicator drum a clear movement. When the indicator is rotating back and forth, if the pencil is held against it the atmospheric line may be drawn. The cock should then be opened and the steam allowed to enter from one end of the cylinder until the indicator has become warmed up. Then the pencil should be held against the drum while the piston takes two or three strokes. A diagram can be taken from the other end of the cylinder on this same card by simply turning the cock over, or this card may be taken out and a new one put in.

Reading an indicator diagram.—To read an indicator for perfect valve setting it is best to compare it with a perfect diagram. It is assumed that in the diagram Fig. 269 the heavy line is the perfect one and those with dotted lines are taken from engines with poorly set valves.

a shows too early compression.
a′ shows too late compression.
b shows excess of lead.
b′ shows insufficient lead.
c shows wire drawing.
s′ shows late release.
s shows early release.

To read an indicator diagram for pressures.—Whenever possible the scales should be divided into parts equivalent to the scale of the spring, i. e., if the spring is 60 pounds to the inch the scales should be divided into 60 parts. Whenever this is not possible a tenths or hundredths scale may be used. The scale shown in Fig.

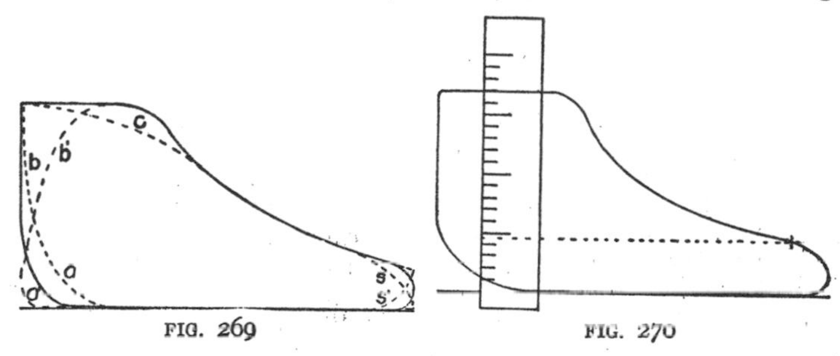

FIG. 269 FIG. 270

270 is a tenths scale, and it now reads 1.7 inches with a 60-pound spring. This gives a steam pressure at that point of

$$1.7 \times 60 = 102.0 \text{ pounds.}$$

If the scale is moved down to the point of release it reads

$$0.45 \times 60 = 27.00 \text{ pounds.}$$

Governors.—The object of a governor is to maintain as nearly as possible a uniform speed of rotation of the engine. When the speed of the engine varies through several revolutions because of variation of load or boiler pressure, the governor will aid in regulating it, but if the variation of speed is confined to a single revolution or a part of a revolution, the variation must be cared for in the flywheel. Since governors for steam engines are attached to the engine, they cannot regulate the speed exactly, for they cannot act until the engine does. In other words, the engine has to commence to slow down before the governor will be affected. It then takes the governor a little time to act, and consequently the engine has quite a chance to vary its speed of rotation. In practice, however, when a slight change of speed takes place, a good governor acts instantly and allows only a very small variation of speed. Governors regulate the speed of an engine in two ways: by varying the steam pressure as it enters the cylinder, and by varying the point of cut-off.

Throttling governors.—Governors which act upon the steam in such a manner as to vary the pressure in the cylinder are known as throttling governors (Fig. 271). In other words, they throttle the steam before it enters the steam chest so there is not enough admitted to fill the space intended for it. Therefore, boiler pressure is not attained, and consequently the steam does not exert as much force upon the piston as when the governor is not acting. As a result, the engine does not do its full capacity of work.

FIG. 271—THROTTLING GOVERNOR

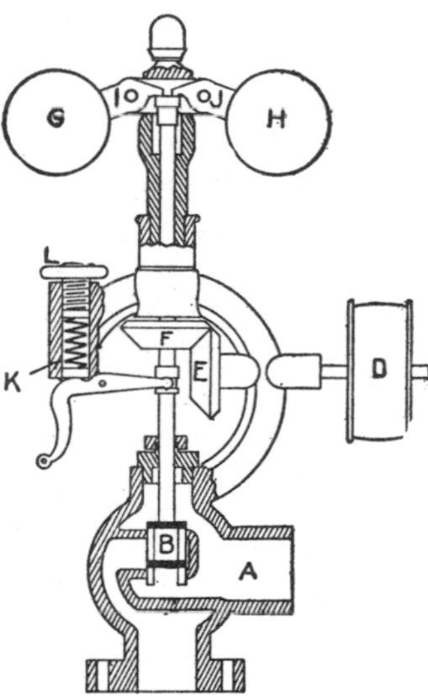

FIG. 272—SECTIONAL VIEW OF THROTTLING GOVERNOR

Principle of the throttling governor. — Fig. 272 is a sectional view of a throttling governor. The governor is generally placed upon the steam chest, and when not in this place it must be as close to it as possible.

Steam enters the governor from the boiler through the pipe *A*. Passing through the governor valve *B*, it enters the steam chest *C*. If the valve *B* is clear up, which is analogous to wide open, the steam passes into the steam chest unmolested as far as pressure is concerned, but if the valve *B* is partly closed the steam is throttled as it passes the valve. Consequently the pressure in the steam chest is not as great as in the steam pipe *A*. From this it is seen that the only requisite for a governor, other than the design of valve *B*, is some device which will raise and lower the valve *B* as the speed of the engine increases or decreases.

The pulley *D* is run by a belt from the engine shaft, and whenever the speed of the engine varies the speed of this pulley also varies. By means of the beveled pinions *E* and *F* the motion is transmitted from the pulley *D* to the governor balls *G* and *H*. With no motion in the pulley *D* these balls hang down, but as soon as the pulley commences to revolve the balls do likewise, and, due to centrifugal force, they commence to rise. When the engine attains its full speed the balls, acting through the arms *I* and *J* and valve rod, should have partly closed the valve. By having the valve partly closed when the engine is running at its normal speed there is opportunity for the valve to be opened when the speed drops. If the engine is not carrying full load it will be inclined to run too fast. This increased speed of the engine causes the governor balls to rise higher and consequently close the valve a trifle. It will be noticed that the governor balls are not only acting on the valve *D*, but are also acting on the spring. Hence if the spring *K* is tightened by screwing down on the hand wheel *L* the engine will have to be running faster before the governor will act. If the hand wheel is loosened, the balls will act more quickly, and consequently the engine cannot attain so high a speed.

If the belt of this governor be taken off, the engine will have to be controlled by the throttle, since there is nothing else to prevent the steam from flowing into the cylinder as fast as the cylinder will take it. If there is no one at hand to control the throttle, the engine will run away. This is the reason why so many engines run away when the governor belt breaks. A great many governors are now equipped with an idle pulley running on the governor belt. This pulley is attached to the throttle in such a manner that when the belt breaks the pulley is free to fall, and by so doing closes the throttle and stops the engine.

Racing.—An engine is said to be racing when its speed of rotation fluctuates badly with a constant load. Racing in nearly all cases is caused by the governor. Either it is not working satisfactorily, or else it is poorly designed. If the valve stem is packed very tight, the engine will have to attain a very high speed before the balls have sufficient force in them to force the valve down. Then when it is down the engine has to slow down entirely too much before the spring will have the energy to force the valve up. An engine will also race if the governor belt is loose and slips, or if the governor is improperly oiled.

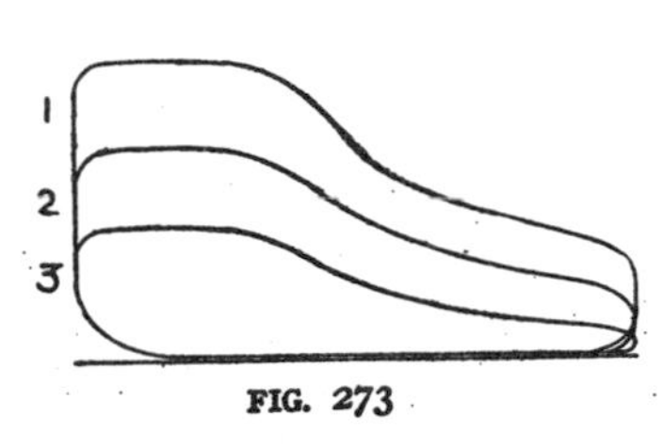

FIG. 273

536. Indicator diagrams from a throttling-governed

engine.—The indicator diagram shows more clearly the effect of throttling the steam of an engine than any description. Fig. 273 shows a diagram taken from an engine; No. 1, with full load; No. 2, with about half load; No. 3, with about quarter load.

In all the diagrams it will be noticed that the points of compression, admission, cut-off, and release remain constant, while the steam and expansion lines vary.

Automatic cut-off governor.—Steam cannot be used as economically under low pressure as under high,

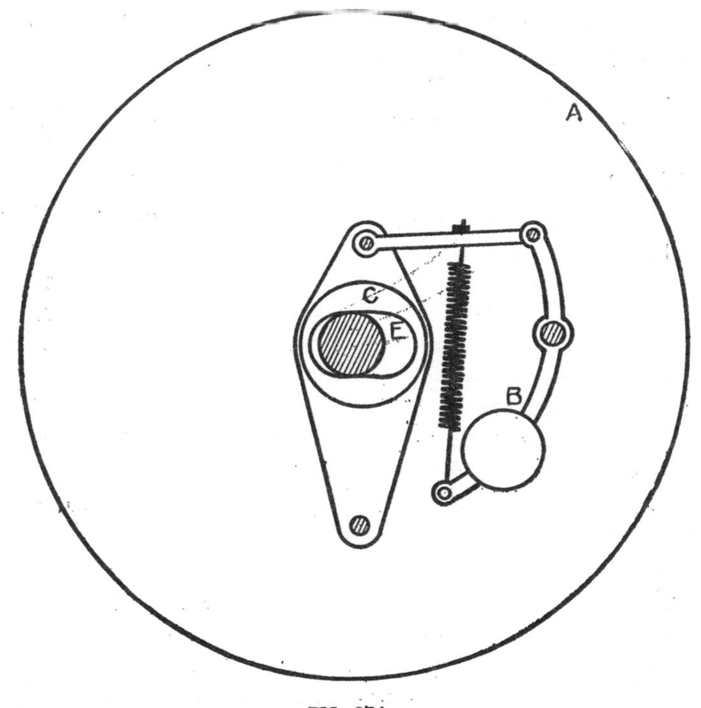

FIG. 274

hence when the steam is throttled down as in No. 3 (Fig. 273) it is not as economical as when used at full pressure. To overcome this loss in throttling-governed engines, automatic cut-off governors have been devised. These governors act in such a manner that they do not throttle the steam as it enters the engine, but change the point of cut-off and by so doing permit steam to enter for a shorter or longer part of the stroke (as the speed of the engine requires) at boiler pressure and allow it to work on expansion. Fig. 274 represents the outline of an automatic cut-off governor. *A* is the flywheel which carries the governor mechanism; *B*, the governing mechanism; *C*, the eccentric sheave; *E*, a slot in the eccentric sheave within which the engine shaft revolves. As the speed of rotation of the engine varies, the weight *B* will move the sheave *C* backward or forward across the engine shaft. This change in the position of the eccentric sheave changes the throw of the eccentric and consequently the point of cut-off of the valve. Fig. 275 shows indicator diagrams with varying loads. No. 1 is overload; No. 2, full load; No. 3, about half load, and No. 4, practically no load. The steam line for all loads is the same, but the point of cut-off varies, thus giving an increased or reduced amount of energy exerted on the piston. Diagram No. 4 shows how the steam has expanded below atmospheric pressure, and when the exhaust port is opened the pressure rises instead of falling. The area below the atmospheric line is then negative work. Instead of working a large engine on as light a

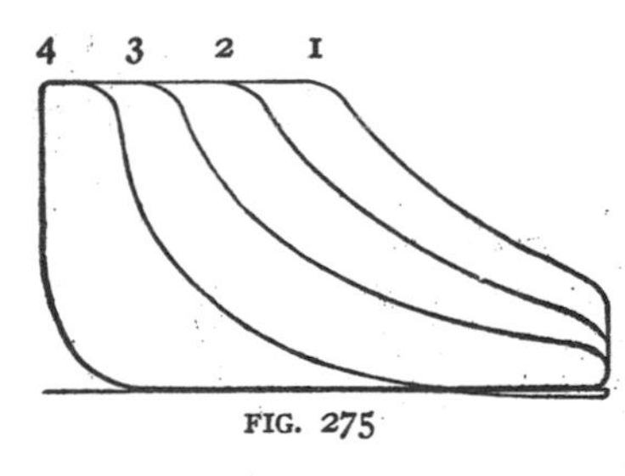

FIG. 275

load as this, much of the time it is more economical to use a smaller engine.

Corliss-governed engines have a great many economical advantages over other types of engines: (1) reduced clearance volume, due to the proximity of the valves to the cylinder; (2) separate valves for steam and exhaust, the steam valves being on top and the exhaust beneath, so there is a free and short passage for the water to leave the cylinder; (3) a wide opening of the steam valve and a very quick closing at cut-off, thus giving a sharp point of cut-off without wire drawing; (4) the valve mechanism permits of independent adjustment of admission and cut-off release and compression. The disadvantages of this engine are that it is of necessity slow speed, and hence to get the required power must be large. This makes the first cost great, not only in the engine itself, but in the material for an engine room.

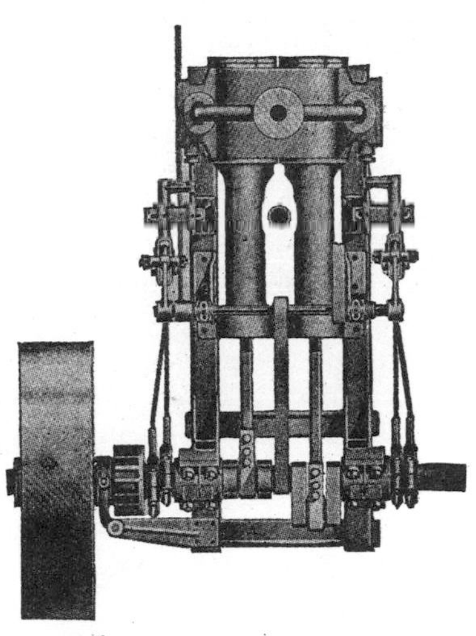

FIG. 276—DOUBLE-CYLINDER ENGINE

A double-cylinder engine (Fig. 276), or a double engine, as it is sometimes called, is an engine which has two cylinders, both of which take the steam directly from the boiler. Both cylinders of a double engine should be connected to the same crank shaft, and their cranks should be at an angle of 90° with each other. The only advantages to be gained from a double-cylinder engine are: (1) being able to start without turning off dead center by hand; (2) being able to start with a heavy load; and (3) being able to move slowly with a heavy load. If the cranks were set in line with each other, these advantages would not be gained. The disadvantages of a double engine are: more moving parts, greater chances for steam to leak about the cylinder and the piston, and more cooling surface, hence greater condensation during the working stroke. Although a double engine is more easily handled than a single one, there are only a few instances, such as plowing and heavy traction work, where its use is recommended for farm work.

Compound engines.—The purpose of compound engines is not to give a greater expansion. This could be accomplished with the low-pressure cylinder and early cut-off. The real purpose is (1) to keep the cylinders as nearly as possible at the temperature of the entering steam, preventing losses by condensation; (2) to reduce the surface exposed to the high-pressure steam to a minimum; (3) to use the high-pressure steam in a small cylinder, hence requiring less material to make it sufficiently strong.

The first cylinder, known as the high-pressure cylinder, expands the steam partly; then the second, or low-pressure, receives it and expands it further. Since the steam as it enters the high-pressure cylinder is under a higher pressure than when it enters the low-pressure cylinder, the latter cylinder must be larger than the former to accommodate the increased volume of the steam. Where steam is expanded in two cylinders the engine is known as a double-expansion compound; where it is expanded in three cylinders it is known as triple-expansion, and in four cylinders it is known as quadruple-

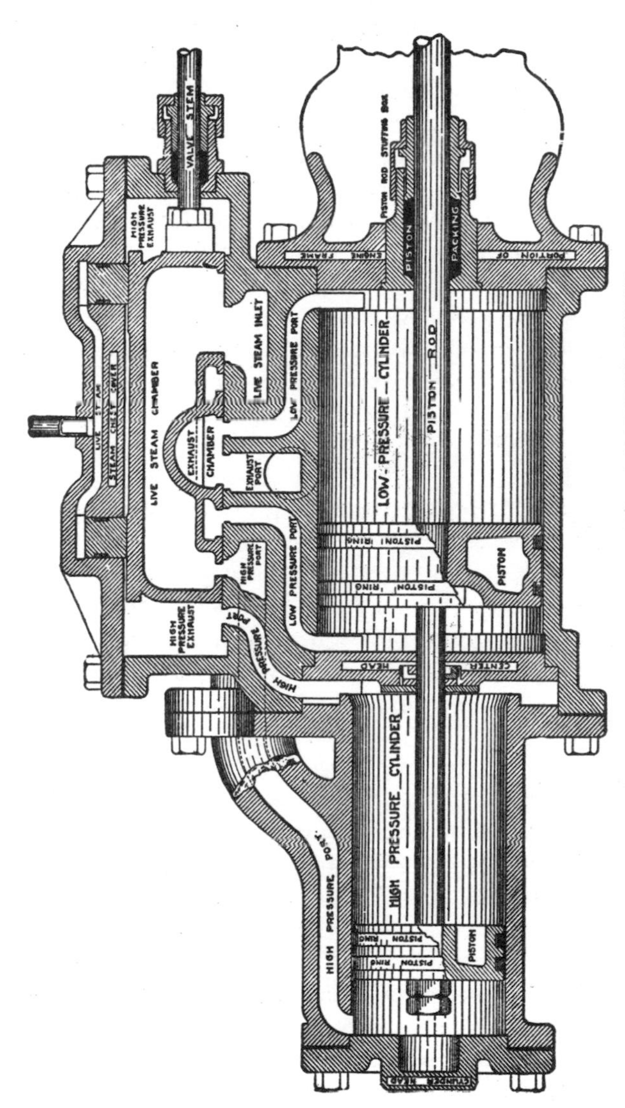

FIG. 277—CYLINDERS OF WOOLF TANDEM-COMPOUND ENGINE

expansion. When one cylinder is in front of the other, the engine is said to be a tandem compound, and when the cylinders are side by side it is said to be a cross-compound engine. Fig. 277 shows the Woolf tandem-compound engine in common use in traction service. The arrows show the direction of the steam as it passes from cylinder to cylinder. Fig. 278 illustrates a cross-compound engine, showing how the steam passes through a superheater as it travels from the high-pressure to the low-pressure cylinder. It also shows the relative sizes of the two cylinders.

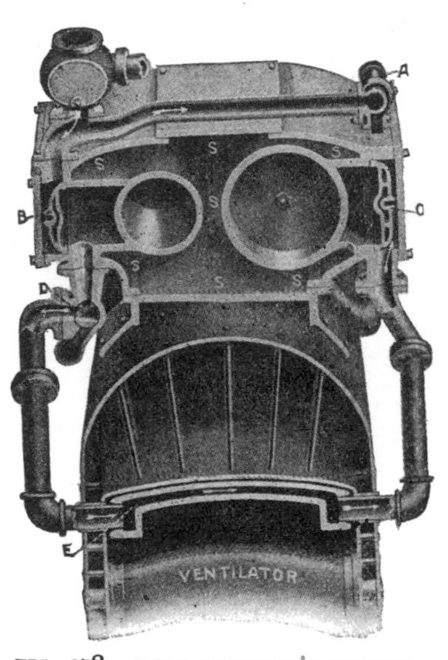

FIG. 278—CROSS-COMPOUND ENGINE

541. **Horse power of steam engines.**—There are three methods of rating steam engines. One method is by the indicated horse power, which is the total work exerted by the steam in the cylinder; the second method is the actual or brake horse power (see Chapter I), which is the actual work delivered from the flywheel of the engine; and the third is the commercial rating.

Commercial rating of steam engines.—The commercial rating of all stationary steam engines is about their actual horse power, but the commercial rating of traction steam engines is far below their actual horse power. This is a custom which originated in the horse power and is to be regretted.

At the time separators were run with horse power they were smaller than they are now and with fewer accessories. At that time 12 horses, by being overworked, would run the separator, but now the separators are larger and are equipped with self-feeders, band cutters, wind stackers, weighers, etc. All of this causes the new separators to run several horse power harder than the old ones. Although the present separators require much more power than the former ones did, competition has kept the rating of the engines down to that of the horse power, while factories are building them much larger. Most traction engines will develop at the brake three times as much power as their rated capacity.

A better way to judge the capacity than by its commercial rating is by the diameter of the cylinder, the length of stroke, and the number of revolutions of the flywheel a minute.

HANDLING AN ENGINE

Starting the engine.—In starting an engine the operator should always see that the cylinder cocks are opened. While the engine has been stopped the steam has condensed and caused considerable water to form in the cylinder, and if there is not some means of letting this out there is danger of injury to the working parts. Even if no water has collected in the cylinder while the engine has been standing, the cylinder walls will be cold and condense the steam as it first enters. It is also well to open the pet cocks from the steam chest and allow the water in there to drain out, and not be carried through the cylinder. The throttle should be only partly opened at first in order to allow the cylinder to become warmed up before full steam is turned on. If full steam is turned on at once there is danger of more water being condensed than the cylinder cocks will carry away. If the engine has a reverse gear, it may be worked back and forth and thus both ends of the cylinder allowed to warm up at once. As soon as the engine has reached its speed and dry steam comes from the cylinder cocks they can be closed and the throttle thrown wide open. The cylinder lubricator and other oil cups can now be started, and if necessary the boiler pump or injector.

Running the engine.—After the engine is once started all bearings should be watched to see that they do not heat. When they get so warm that the hand cannot be borne on them the engine should be stopped and the bearings loosened. If the engine runs properly, all repairs that can be made while the engine is in motion should be attended to: the oil supply looked after, oil cups kept full, etc.

Stopping the engine.—To stop the engine the throttle should be closed and the cylinder cocks then opened. The throttle may be closed quickly without injury to the boiler or the cylinder, providing there is plenty of water in the boiler. Close all lubricators. The cylinder cocks should be left open until after the engine starts again. If the engine is stopped for only a short interval, the cylinder walls will cool off so little that the engine can be quickly started. It is not well, however, to start the engine into full speed at once. This throws too much strain on the working parts.

Leaks.—Engines should be occasionally tested for steam leaks past the valve or the piston. The easiest and surest method to do this is to use the indicator, but wherever this is not possible the valve can be tested by placing it in its central position and turning on steam.

If there is any leak, condensed steam will flow from the cylinder cocks.

If the valve is tight, leaks past the piston may be found by blocking the crosshead so as to hold the piston in one place, then turning steam into one end of the cylinder. If water comes from the cylinder cock in the other end, steam is leaking past the piston.

It is well to make this test for both ends of the cylinder and with the piston in two or three positions. Sometimes the piston rings will allow the steam to pass one way and not the other. Often there are irregularities in the inside surface of the cylinder, and steam will leak past the piston when it is in one part of the stroke and not in another. Although a small leak may not appear very important, all the steam which leaks past the valve or the piston passes off into the exhaust without doing work. When the valve leaks it should be taken out and scraped to a fit. If the piston leaks, new rings should be put in, and if it continues to do so, the cylinder should be rebored.

Packing.—There are two classes of packing, piston packing and sheet or gasket packing. The former is to be used where moving parts are to be packed, such as piston and valve rods. It generally consists of some sort of wicking, such as candle wicking, asbestos wicking, hemp wicking, or patent wicking. Candle wicking and hemp are good all-purpose packings, but should not be left in the packing box too long, as they will become hard and cut the rod. Asbestos wicking is good packing for all purposes but pump rods. It does not get hard like hemp or candle wicking, but the water on a pump rod soon washes it out. Patented packings will last longer and not get hard like the common packings. Gasket or sheet packings are used on pipe fittings, manholes, and handholes, where there is no motion. Such packing should be just thick enough to cover the uneven surfaces and no more.

Pounding.—An engine which pounds is generally loose or worn, and if permitted to continue pounding will gradually become worse. The wrist and crank bearings are those most likely to pound. Nevertheless, there are so many other places where the engine will pound that it is well to look not only at these points, but at others. An experienced engineer will have no trouble in detecting the exact place, but a new man should work cautiously. He should block the crosshead, and then turn the flywheel backward and forward an inch or two. This will tell whether the pound is in the flywheel, main bearings, crank pin or wrist pin. This will not tell, however, if the pound is in the governor belt pulley, or guides. A new engineer should not try to take out all of the pound at once; only take up the slack a trifle at a time until it is all removed. It is better to run a box too loose and have it pound than too tight and have it cut. An engine may also be loose in the eccentric and valve, and cause pounding, or sometimes it will pound when out of line. In the former case a little tightening will remove the pound; not too much, however, or the eccentric may cut or the valve bind. If the engineer thinks the shaft is out of line, he can detect it by taking the front half of the crank bearing off the connecting rod, and then by inspection see if the connecting rod freely rests in its position in the crank pin in all parts of the stroke of the piston.

Bearings.—All important boxes and those which are likely to wear should be made in halves with liners between the halves. This permits of taking up the wear, without requiring a new bearing. The ideal bearing is a perfectly round hole with a pin fitting it just close enough to allow a film of oil between the hole and the pin. The closer a bearing can be made to conform to this the better.

As a bearing wears, a thick liner should be taken out and a thinner one inserted. Never take out a thick liner and then only partly draw up the boxes. This makes a loose bearing and will cause trouble.

Lubrication.—Since the cylinder of the engine is always hot when running, oil is required which will stand higher temperatures than the oils for bearings. This oil is generally known as cylinder oil. As a rule, it is a heavier and blacker oil than is generally used for lubrication. It is of such a nature that it will stand the heat in the steam chest and the cylinder. Ordinary lubricating oil would be decomposed by the heat. The oil used for bearings, such as crank, eccentric, wrist pin, etc., is of a lighter nature and is a good grade of common lubricating oil. A new engine requires more oil than an old one, and a cylinder when priming or foaming requires more oil than when running regularly. The amount of oil to use can be determined only by experience; it is better to get too much than not enough. A good method of determining the amount of oil for the cylinder is to keep track for a minute of the number of drops which pass through the lubricator; then take the cylinder head off and see if the walls are bright and shiny and feel oily; if so, the cylinder is getting enough oil. For bearings and other places, the number of drops a minute should be determined, and then the bearings watched to see if they heat and if there is an excess of oil running off.

Lubricators.—Owing to the pressure in the steam chest of an engine, some device has to be employed which will force the oil into the steam pipe against this pressure. There are several makes of lubricators on the market. Fig. 279 shows the principle of nearly all of them. This lubricator is so arranged that the steam condenses in the small pipe of the lubricator and forms a greater pressure on one side of the oil than on the other. This forces the oil from the valve to the steam pipe. To fill the lubricator, the cocks from the steam pipe should be shut off so no pressure can be let in; then the small cock at the bottom of the lubricator should be opened and the condensed water let out. When oil commences to come instead of water the lubricator has been drained enough. The cock can then be closed and the cap on top taken off and the oil poured in. Several makes of oil pumps now on the market are to take the place of the lubricator. They are actuated by a lever and arm from the crosshead. These pumps are more positive than lubricators in their action and not as likely to fail to operate. The only defect in this form of lubrication is that very few pumps have a sight feed or a glass which will tell how much oil is in the vessel that contains it; thus it is hard to tell whether the pump is full or empty.

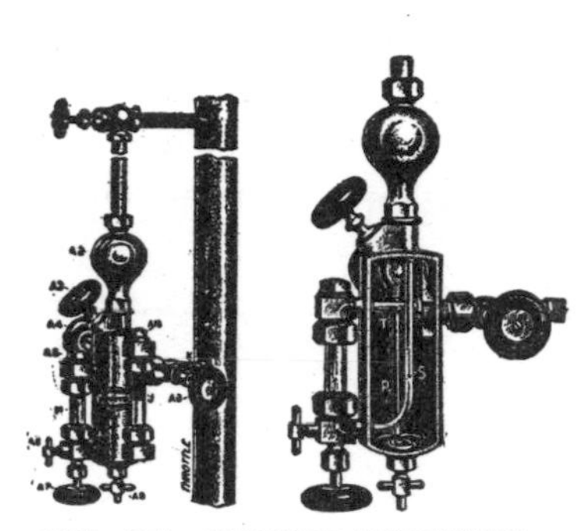
FIG. 279—CYLINDER LUBRICATOR

CHAPTER V

GAS, OIL AND ALCOHOL ENGINES

Internal-combustion engine. — The gasoline engine is of the type known as the internal-combustion engine. Others of this type are the gas engine, the hydrocarbon engine, the kerosene engine, the oil engine, and the distillate engine.

In the steam engine combustion takes place in the furnace; the heat is diffused through the boiler, generating steam; this steam is then transferred by means of pipes to the engine. Through all these operations a great deal of heat energy is lost by radiation. In the internal-combustion engine the fuel is put under high pressure by the inward movement of the piston. While in this condition it is ignited; the consequent burning causes a very great expansive force, and this force, acting directly upon the moving parts of the engine, gives very little opportunity for radiation.

The principle of all internal-combustion engines is the same, so in this chapter the gasoline engine will be used as a basis of discussion. The gas and the gasoline engine are so nearly identical that they may be treated in the same manner.

Early development.—At first the development of this engine was very slow. Huyghens in 1680 proposed the use of gunpowder. Papin in 1690 continued the experiments, but without success. Their plan was to explode the powder in an enclosed vessel, forcing the air out through check valves, thus producing a partial vacuum, causing the piston to descend by atmospheric pressure and gravity. The few experimenters who took up the work continued in this line with more or less improvements until 1860, when Lenoir brought out the first really practical engine. This was very similar to a high-pressure steam engine using gas and air. Among these early experimenters, those who seem most prominent are Barnet, in 1838, inventor of flame ignition and compression, and Barsanti and Matteusee, who, in 1857, produced the free piston.

Later development.—Million gave the first clear ideas of the advantages of compression, and M. Beau de Rochas went further and produced a theory analogous to our present type. In 1867 Messrs. Otto and Langdon produced a free piston engine which superseded all previous efforts, but it was left to Mr. Otto to put into practice

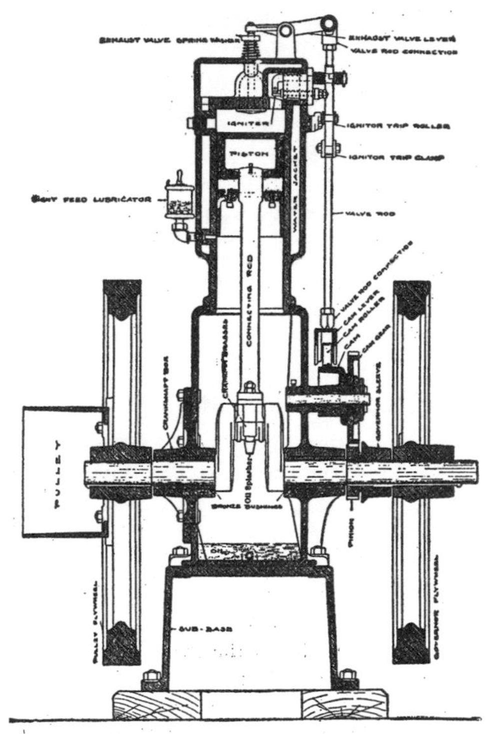

FIG. 280—PARTS OF GASOLINE ENGINE

in 1876 the first engine of commercial value. All present-day types work on the same principle as Otto's, but under fewer practical difficulties.

Types of gasoline engines.—Otto was the first to put into practice the idea of compressing the gas and air mixture before igniting. This gave rise to the name of Otto cycle, which is now used in all engines. Compression is one of the important things which determine the economy of the engine; theoretically, the efficiency of the engine depends upon the compression pressure. However, it is not possible to increase the compression pressure indefinitely because the charge pre-ignites and causes the engine to pound. It is desirable, however, to use as high a compression as possible.

As stated before, practically all the engines used today are designed to follow the Otto cycle. However, they are divided into two distinct types, four-cycle engines and two-cycle engines.

Four-cycle engines.—The term cycle is applied to the entire operation of converting heat into mechanical energy. In the four-cycle engine four strokes of the piston

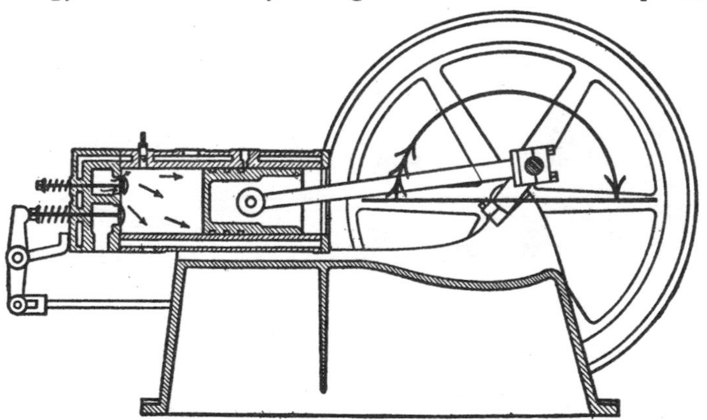

FIG. 281—SUCTION STROKE OF FOUR-CYCLE ENGINE

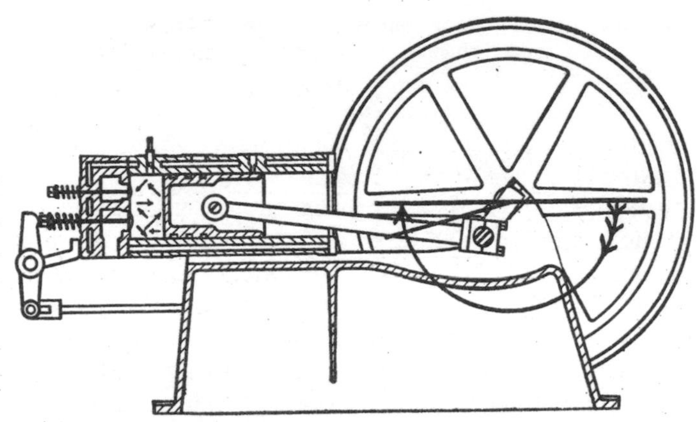

FIG. 282—COMPRESSION AND IGNITION STROKE OF FOUR-CYCLE ENGINE

or two complete revolutions of the crank are necessary to complete this cycle, hence the name four-cycle. These strokes may be enumerated as follows: The piston makes one forward stroke, drawing into the cylinder through the inlet a charge of fuel and air. This is called suction (Fig. 281). A second stroke compresses this charge

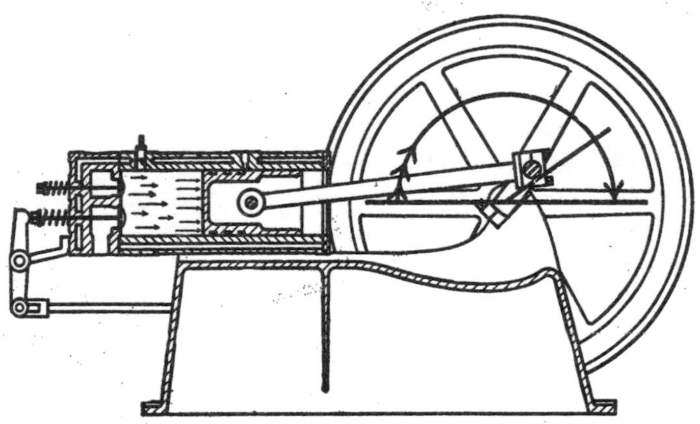

FIG. 283—EXPANSION AND RELEASE STROKE OF FOUR-CYCLE ENGINE

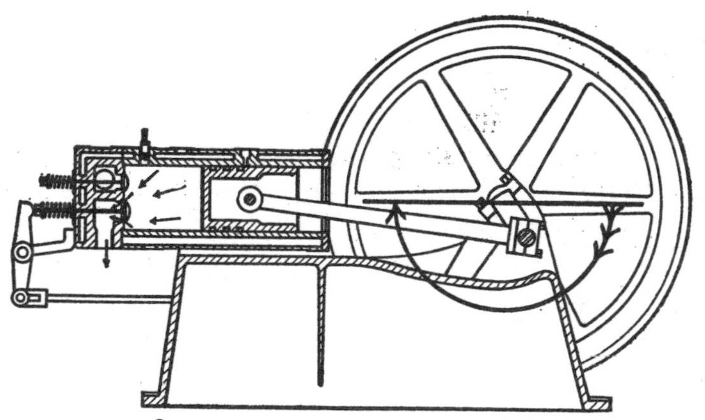

FIG. 284—EXHAUST STROKE OF FOUR-CYCLE ENGINE

into the clearance space of the cylinder (Fig. 282). This stroke is called compression. Just before the crank passes dead center the charge is ignited. Owing to the heat released, the gases expand, and this expansion of gases acts upon the piston, driving it forward during the third stroke, which is called expansion (Fig. 283). This stroke is the only working stroke of the cycle. During the fourth stroke the exhaust valve is forced open by mechanical means and the piston crowds the burned gases out. This stroke is called exhaust (Fig. 284).

The two-cycle engine completes the cycle in two strokes of the piston and from this fact derives its name. In this type of engine there must be, besides the cylinder, a compression chamber, which may be separate, which may be the crank case enclosed, or which may be the front end of the cylinder. To illustrate the cycle in this type of engine, the enclosed crank case type is used. That is, the cylinder and the crank case are both gas tight and practically in one piece. However, the two chambers are separated by the piston. Let the piston be at the crank end of the cylinder, then start it up. This action tends to produce a vacuum in the crank case, but instead of doing so the charge rushes in through the check valve *A* (Fig. 285) and fills the space. Now start the piston down again, compressing the charge in the crank case (Fig. 286) until the piston has opened the inlet port, when the charge rushes from that end of the cylinder up

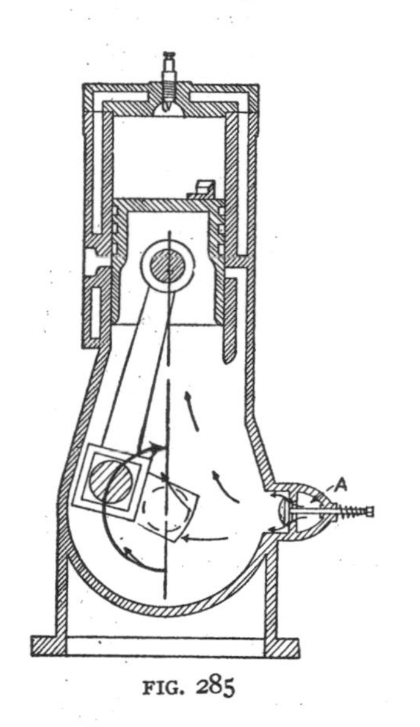

FIG. 285

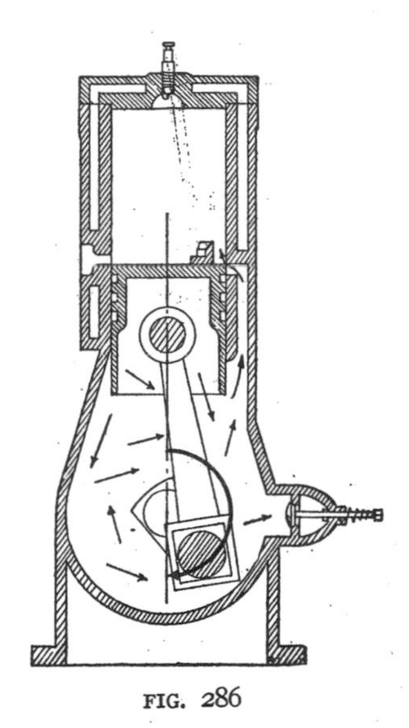

FIG. 286

into the other. As the piston starts back again (Fig. 287) it closes the openings and compresses the charge now in the head end. At the same time it is doing this a new charge is being drawn into the crank case. When the piston reaches the upper end of the stroke explosion takes place and the expansion forces the piston down (Fig. 288), compressing the charge in the crank case and expanding the one in the cylinder. When the piston has passed the port openings the burnt gas rushes out through the exhaust port and the new charge comes in through the inlet port. Thus we see that when the piston is compressing a charge in the cylinder a new charge is being taken into the crank case, and when the charge in the cylinder is expanding the gas in the crank case is being compressed.

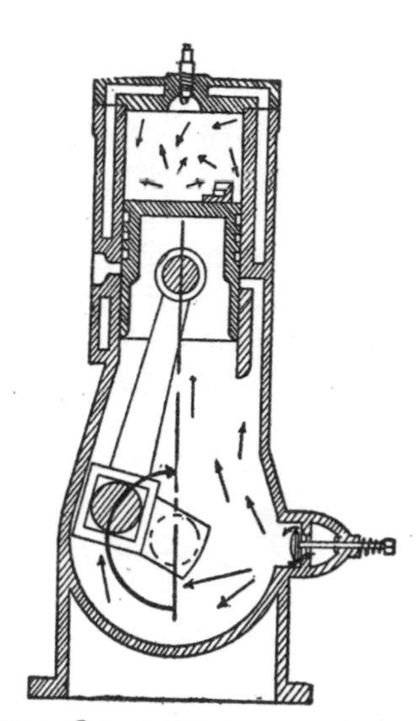

FIG. 287—SUCTION, COMPRESSION AND IGNITION STROKE OF TWO-CYCLE ENGINE

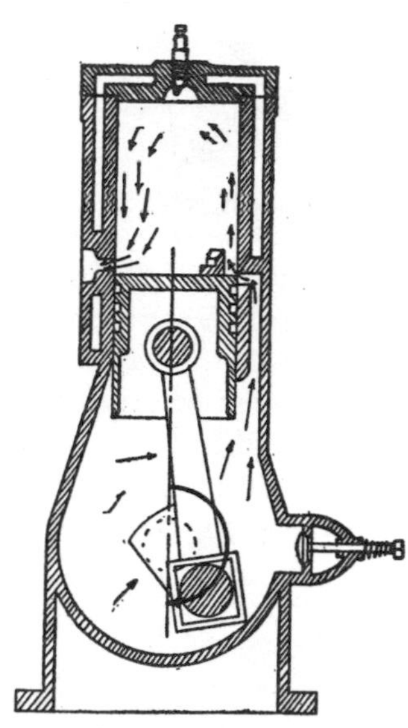

FIG. 288—EXPANSION, EXHAUST AND INLET STROKE OF TWO-CYCLE ENGINE

Construction. — (Only the four-cycle engine will be considered hereafter.) The parts of a gasoline engine necessary to be examined for proper construction are: cylinder head, cylinder, base, piston, and piston rings, connecting rod, crank shaft, flywheels, valves, governor, carburetor, ignitor, and cooling device.

FIG. 289—CYLINDER HEAD

Cylinder head.—The cylinder head (Fig. 289) should have a device for cooling. If water is used for this, the inside of the head should be at least ⅜ inch thick for a 5-inch cylinder and the water jacket not less than ¼ inch. These dimensions increase with the size of the engine.

Cylinder.—The cylinder (Fig. 290) should be bored perfectly smooth and round, and should be free from all flaws and imperfections. It may have the same thickness of castings as the head.

Base.—The base (Fig. 291) should be designed to carry the cylinder, engine frame, and flywheels in a well-balanced condition.

FIG. 290—CYLINDER

Piston.—The piston (Fig. 292) is one of the important parts of the engine. It should be of good length to carry itself without binding. The piston pin should be near the middle and as long and as large as possible. In

FIG. 291—BASE

small engines the piston should be about 1/200 of an inch smaller than the cylinder, and in larger sizes it should be about 1/32 of an inch smaller. The space on the head end of the piston beyond the last ring should be about 1/16 of an inch less in diameter than the rest of the piston.

Piston rings.—The number of rings (Fig. 292) varies from three in cylinders of 5-inch diameter and less up to eight in 20-inch cylinders. If the engine is of the

FIG. 292—PISTON AND RINGS

vertical type, there should be a ring at the lower end of the piston. This ring will prevent "oil pumping." The rings should break joints, and if one edge fits closer to the cylinder than the other, the close-fitting edge should be toward the explosion end. All rings should be crescent-shaped. This causes an equal pressure all around.

The connecting rod.—The connecting rod (Fig. 293)

FIG. 293—CONNECTING ROD

should be of forged steel and of the right weight to carry the load. A simple bearing is sufficient at the wrist pin, but at the crank end the boxings should be held in place by means of two bolts. All parts should be in perfect alignment.

Crank shaft (Fig. 294).—It is essential that the crank shaft be heavy enough to withstand the sudden shocks which come to it from the explosions, and it should also carry the heavy flywheels without springing. The bearings should be long and in perfect alignment. Their line of centers must be exactly at right angles with the cylinder. A good way to detect a weak crank shaft is to notice whether the flywheels wobble at each explosion of the engine.

Flywheels (Fig. 295).—These are necessarily heavy and massive, but not necessarily ungainly in appearance. Loganecker says: "At a medium speed, which may be based on about 225 revolutions for 25 H.P. to 375 for 2 H.P., 100 pounds to the horse power will not be very far out of the way. The di-

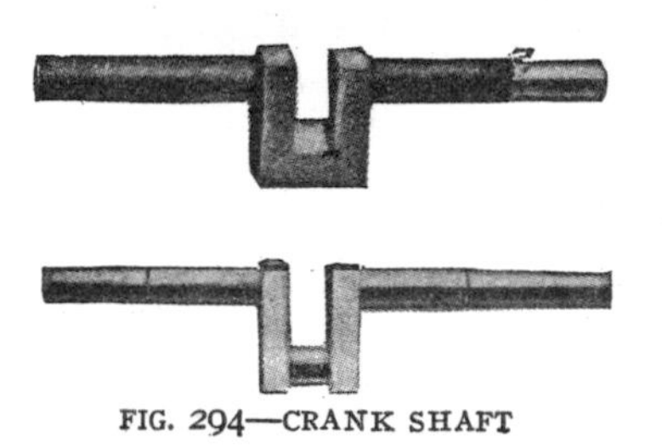

FIG. 294—CRANK SHAFT

ameter may range from 28 inches on the small engine to 60 inches on the larger size." The above weight is to be divided between the two wheels.

FIG. 295—FLYWHEELS

Valves (Fig. 296).—It makes no great difference where the valves are located, just so they are close to and connected to the clearance space. A good rule to follow for size is: Inlet valve diameter should be five-sixteenths diameter of the cylinder, and the exhaust valve about seven-sixteenths.

Governors.—There are two types of gasoline engine governors in general use. These are the **throttling** governor and the **hit-or-miss** governor.

Throttling governors vary the amount of gasoline mixture admitted to the cylinder. Before the engine has reached its normal speed, or when it is carrying a full load, each charge is a full charge, with as near a perfect mixture as possible. Consequently the normal compression pressure for that engine is attained and the engine does its work with its greatest economy. When the engine is doing only a part of its rated capacity of work, the throttle acts. This reduces the volume of mixture which enters the cylinder, but the space within the cylin-

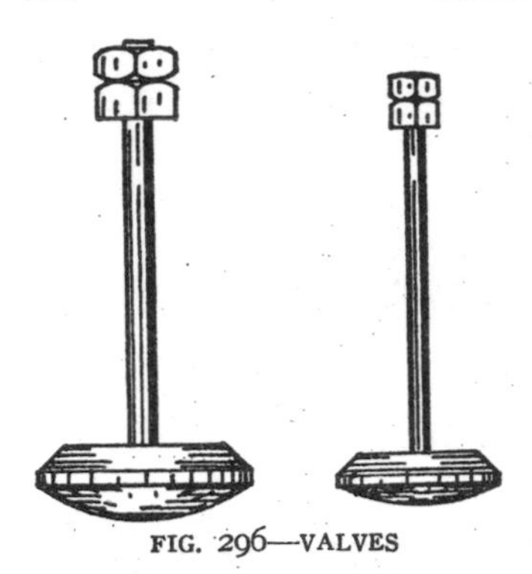
FIG. 296—VALVES

der to be filled is the same; consequently the compression pressure is not as great as it should be and the engine is not economical with fuel. Often the load in an engine is small enough for the charge to be throttled down until of such small volume as not to ignite, but simply pass off to the exhaust unburned. Throttling-governor gasoline engines are not as economical with a variable load as the hit-or-miss type of governed engines. However, their motion is much more steady, and often the matter of economy is waived in order to secure the greater uniformity of speed.

Hit-or-miss type of governor.—In the hit-or-miss type of governor the amount of mixture drawn in for an explosion remains at all times a constant, and the governing is accomplished by cutting out all admissions while the engine is running faster than normal speed. This method of governing is usually accomplished by holding the exhaust valve open and the inlet valve closed until the engine falls a trifle below speed, when the exhaust valve closes and new charges are taken into the cylinder. Fig. 297 shows the manner in which this style of governing is accomplished. When the speed of the engine is above normal the governor sleeve *C*, which is in the crank shaft, is drawn out, and, acting on the detent roller *D*, throws the detent lever *E* down so it becomes engaged in the hook-up stop *F*. The hook-up stop *F*, being connected to the exhaust valve rod *H*, holds the exhaust valve open. By reference to Fig. 298 it will be seen how the inlet valve is held closed. There are three general methods of using weights to accomplish hit-or-miss gov-

FIG. 297—MECHANISM FOR HIT-OR-MISS GOVERNOR

erning as explained above. They are: By having weights in the flywheels (Fig. 299); by having weights in a special shaft (Fig. 300); and by means of an inertia weight

FIG. 298—INLET VALVE LOCK

FIG. 299—FLYWHEEL GOVERNOR

FIG. 300—BALL GOVERNOR

(Fig. 301). The latter type of governor works on the principle that when the engine is running at normal speed the weight does not get enough throw to cause the detent to catch in the hook-up stop, but when the speed is increased above normal the weight is thrown far enough to accomplish this.

Carburetor.—Before gasoline can be used in an engine for fuel it must be converted into a vapor or into a gas. This process of converting the liquid into a gas is called **carburetion,** and the device by which it is accomplished is called the carburetor. It is by means of the carburetor that a proper mixture of gasoline and air is made for combustion in the cylinder. A proper mixture is one of the important functions of successful gasoline engine operation.

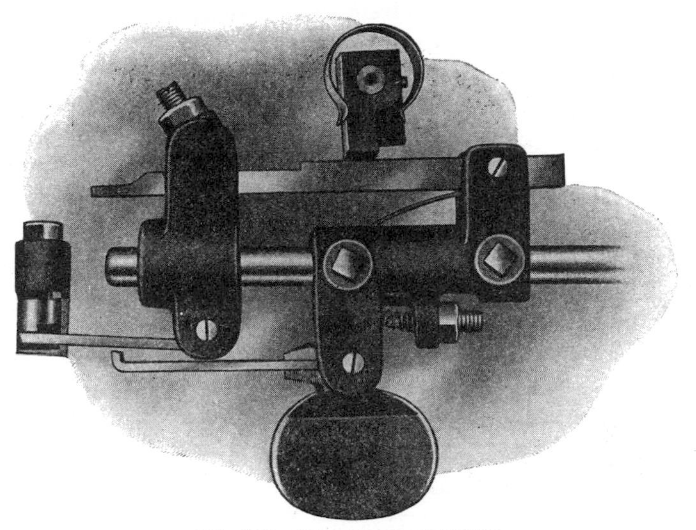

FIG. 301—PENDULUM GOVERNOR

When the liquid gasoline is converted into a vapor its volume is increased about 1,500 times. To make a strong explosive mixture the vapor must be diluted from 8 to 13 times the volume of the air; the air in this case supplying the oxygen. Thus we see that the volume of gasoline to the volume of air used is in the proportion of about 1 to 12,000 or to 19,500. It follows that the carburetor must necessarily be a very delicate arrangement. An engine will not run satisfactorily unless the mixture is very nearly correct. There is a multitude of surface, wick, gauze, spray, atomizing and float-feed carburetors and generator valves on

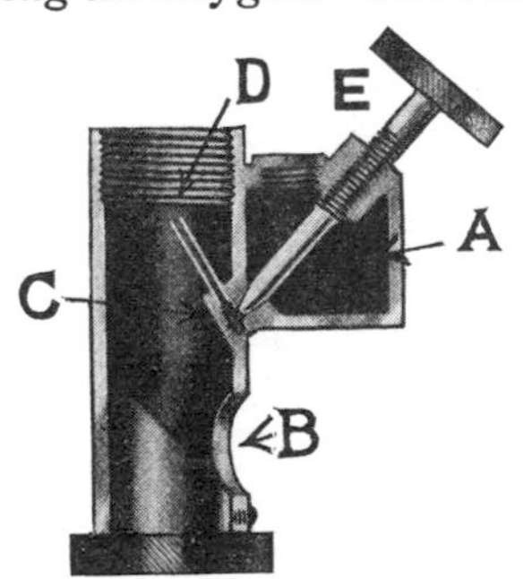

FIG. 302—PRINCIPLE OF THE CARBURETOR

the market. Practically all of these devices depend upon the liquid fuel being caught by the incoming air and atomized on its way to the cylinder. Fig. 302 illustrates the principle of the carburetor. Gasoline flows into the chamber *A*; air enters at *B* and passes through the chamber *C* into the engine cylinders. As the air passes the tube *D* it takes up the charge of gasoline which has been admitted through the needle valve *E*, and carries it on into the engine with itself. Since the air passes the tube *D* at a velocity of about 6,000 feet a minute, it immediately atomizes the gasoline and forms it into a gas. Fig. 303 is a commercial carburetor wherein the gasoline is kept at a constant level in the reservoir. Fig. 304 shows a float-feed carburetor, the principle of which is illustrated in Fig. 305. As fast as gasoline is taken from the tube *A* the float *B* drops and more gasoline enters the reservoir *C*.

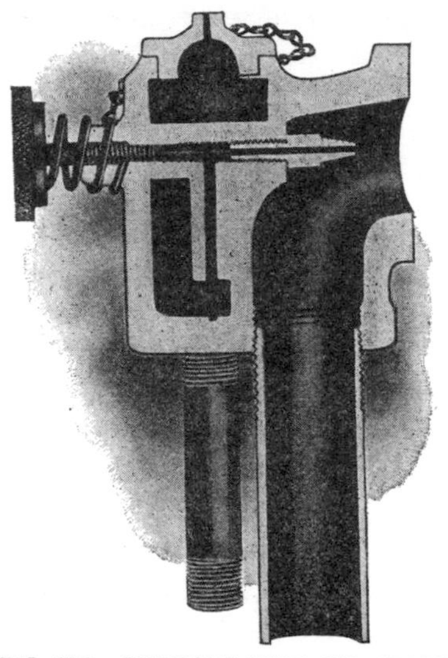

FIG. 303—CONSTANT-LEVEL CARBURETOR

The charge of gasoline taken into the engine each time is so small that the amount can be regulated only by a needle valve. Such valves as are used about the pump are far too large. It is also due to this minuteness of the charge that the gasoline has to be kept at a constant level in the reservoir of the carburetors. For instance, if the carburetor illustrated in Fig. 303 has no overflow, but the attendant endeavors to regulate the amount of gasoline in the reservoir by means of the valve in the feed pipe, he will set his valve so that the engine runs well under a full load, but when the load becomes less fewer charges will be drawn in and the pump will throw the same amount of gasoline. Consequently the reservoir will fill so full that when the engine does take a charge there will be so much gasoline in it that there will not be complete combustion, and as a result the explosion will be weak and the exhaust gas will be black smoke. The carburetor should be near the cylinder to enable the mixture to be easily controlled.

FIG. 304—FLOAT FEED CARBURETOR

Ignitors.—There are two general types of ignitors,

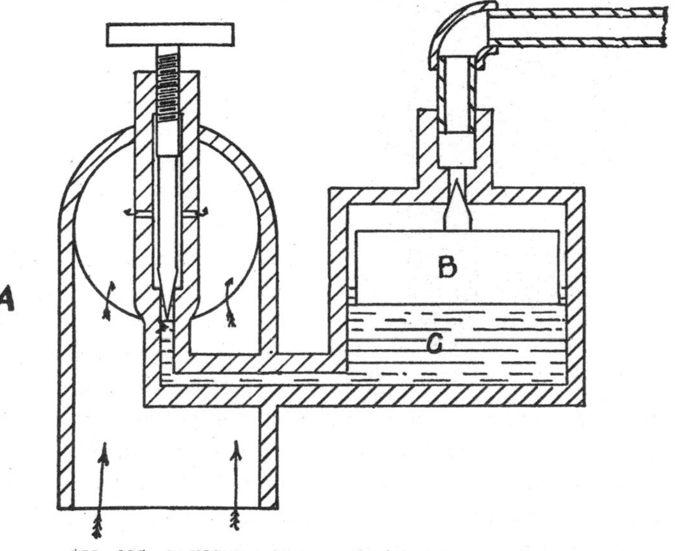

FIG. 305—PRINCIPLE OF THE FLOAT-FEED CARBURETOR

the **hot tube** and the **electric spark.** The latter type, which is most popular in America at present, may be divided into two classes, the **contact spark** and the **jump spark.**

Contact spark (Fig. 306).—It has been noticed that when a break is made in an electric circuit a spark takes place, and it is upon this principle that the contact gasoline ignitor depends. The charges of the fuel mixture are ignited by causing this break to be made inside of the cylinder. The quicker this break is made, the more pronounced the spark. The spark is always made larger and more pronounced by including in the circuit a spark coil.

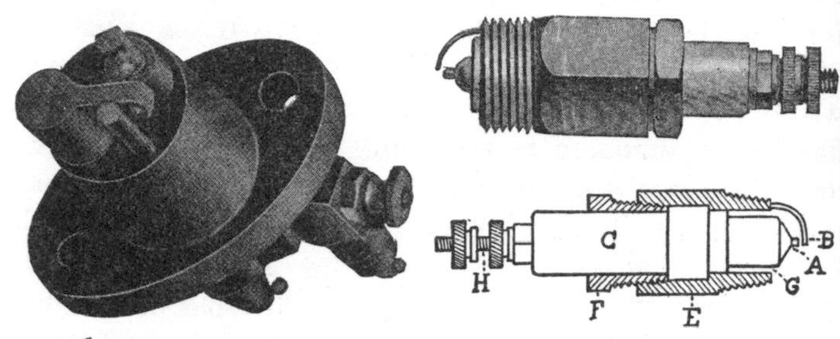

FIG. 306—CONTACT-SPARK IGNITOR   FIG. 307—JUMP-SPARK IGNITOR

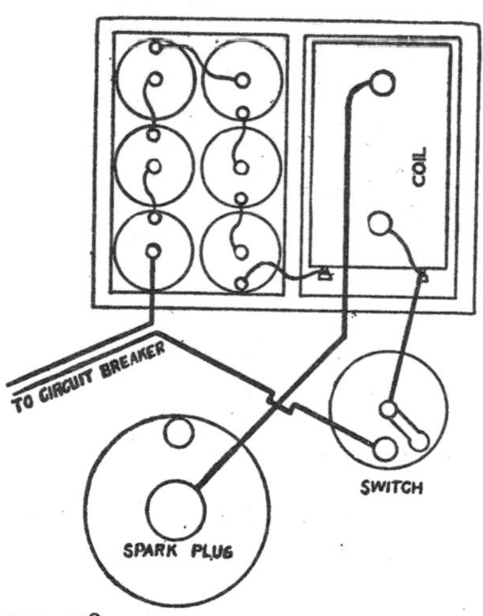

FIG. 308—WIRING SYSTEM FOR JUMP-SPARK IGNITION

Jump spark.—The jump-spark ignitor (Fig. 307) has within the cylinder two points insulated from each other and separated by a very short distance. It differs from the contact-spark circuit in that there must be an induction coil. This coil requires a primary current leading to it from the batteries, and a secondary current leading to the spark points. This latter current has the characteristic of jumping from one point to the other in the form of a spark, thus igniting the charge in the engine (Fig. 308).

Batteries.—In the majority of cases the currents for electric ignitors are furnished by batteries composed of either dry or wet cells. It is very difficult to determine without the aid of proper instruments when a battery has been exhausted to the point where it does not furnish sufficient current. Upon trying an exhausted battery out, it will in all cases give a satisfactory spark. This is due to the fact that batteries when exhausted tend to recover slightly during the rest and are able to furnish current for a few ignitions. Upon starting an engine with an exhausted battery, a few ignitions will take place satisfactorily, but later it will miss fire, due to the weakness of the battery. Often when a battery is becoming run down and the engine is still running the latter will take in several charges, but no explosion will result; then there is an explosion and a great report from the exhaust. This is because the explosion in the engine ignites the unexploded charges which have previously passed through into the exhaust chamber.

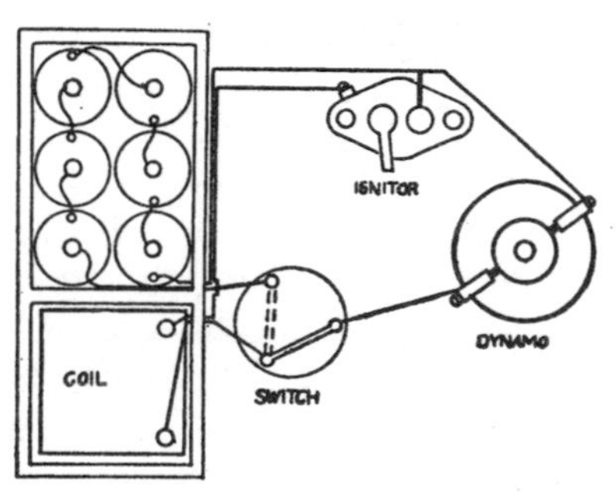

FIG. 309—WIRING SYSTEM FOR BATTERIES AND DYNAMOS

Dynamos and magnetos.—Since the battery is expensive and short lived, other provisions are made for supplying electric currents. One of the most satisfactory of these is by connecting the engine to a form of magneto or dynamo. The amount of power needed to drive a dynamo is exceedingly small, but at all times sufficient current is provided to give reliable ignition. A magneto differs from a dynamo in that the pole pieces of the magnetos are made of permanent magnets, while those of the dynamo are electromagnets.

It is often easier to start an engine with a magneto than with a dynamo. However, after speed is reached, the dynamo, as a rule, is a little more satisfactory than the magneto. These small dynamos are usually provided with a self-governing device which will regulate the speed and in this way obtain the proper voltage foı ignition.

Cooling of gasoline engines.—There are three methods of carrying the excess heat away from the gasoline engine cylinder, namely: (1) air cooled; (2) water cooled; and (3) oil cooled.

The air-cooled engine (Fig. 310) is provided with ribs or flanges extending from the cylinder, which gives up a certain amount of heat to the air. This may be assisted by a draft of air blown upon the cylinder by a fan, bringing more air in contact with the flanges. Air-cooled engines are necessarily of small units, but where the engine is small and exposed to freezing weather it is preferable to any other.

Water-cooled engines are the type in most general use, and water is perhaps the best means of carrying the excess heat from the cylinder. There are three general plans in use for cooling with water. One is to have a large tank sitting near and connected to the engine (Fig.

FIG. 310—AIR-COOLED ENGINE

315). One connection should be from the lowest part of the water jacket to the lower part of the tank; the other should be from the upper part of the water jacket to the top of the tank. The heat from the engine causes circula-

FIG. 311—CIRCULATING PUMP SYSTEM OF COOLING

tion similar to that in a boiler. Another plan (Fig. 311) is to provide some way for the water to fall through the air and thus cool itself by evaporation. In this plan a circulating pump is necessary. The third method is to allow a stream of water to run continually through the engine. The first way is the most economical and possibly the most satisfactory where there is plenty of room and no danger of frost. The second method is coming into general use because it takes less space and does not require so much water at once. All late portable engines are equipped with this device for cooling. For stationary engines and where the amount of water used may be unlimited, the constant-flow method is considered the best, since by this means the water can be drained from the jacket every time the engine is shut down, and turned in again upon starting, and thus avoid the danger of freezing.

Open-jacket cooling.—Engines are now coming upon the market which have the open-jacket method of cooling. The casting for the water jacket is extended so it forms a reservoir upon the top of the engine (Fig. 311*a*). This reservoir is open at the top and holds but a few gallons of water. As the engine heats, the water is allowed to boil and evaporate. Since there is only a pailful or so of water in the engine, it is but a small matter to drain the engine and then refill in cold weather.

FIG. 311A—OPEN-JACKET SYSTEM OF COOLING

Oil cooling system.—By having a radiator and circulating pump, oil is used for cooling where engines are exposed to freezing temperature.

Often chemicals are used in water to prevent freezing. Calcium chloride is the most common of these. The proportions generally used are 5 pounds of the chemical to 10 gallons of water. Whenever possible, the use of chemicals should be avoided; they attack either the tank or the engine castings.

Gasoline engine indicator diagram.—The highest pressure obtained in the average steam engine cylinder rarely exceeds 175 pounds to the square inch. In gasoline engines the average maximum pressure is about 300 pounds a square inch, and it often exceeds 400 pounds a square inch. Since the pressures are so high in the gasoline engine, either the spring has to be made stiffer in the indicator or else the piston made smaller. Either method is utilized with success in indicator work. All parts of the gasoline engine indicator, excepting the spring, are the same as those of the steam engine indicator. In the steam engine the working fluid is admitted to the cylinder ready to perform its work on the piston. In the gasoline engine this is not the case. The working fluid enters the cylinder in the form of a gasoline fuel, which has to be compressed and burned before it is ready for use. Since these operations take place in the engine cylinder, the gasoline engine indicator diagram is different from that of the steam engine. Fig. 312 is a typical gasoline engine indicator diagram and can be followed out thus: *XY* is the atmospheric line; *ABC* is the line produced by the suction stroke of the piston; *CDE* is the compression line; *E* is the point of ignition; *EFG* is the line produced by the increase in pressure as the gas burns; *GHI* is the expansion stroke line; *I* is the point of release; *IC* is the exhaust line and *CJA* is the exhaust stroke line. If the inlet valve is opened automatically the suction stroke line falls far below atmospheric pressure, but if it is opened mechanically the line *ABC* will fall only a short distance below the line *XY*. The indicator diagram shows as clearly what is the matter with a gasoline engine as it does with a steam engine. Fig. 313 shows cards from engines where ignition is too late; Fig. 314, cards which indicate too early ignition.

Losses in a gasoline engine.—If it were possible to utilize all the heat in the fuel in a charge of gasoline, there would be no more economical method of producing power, but the mechanical difficulties which have to be overcome are so great that only about 25 per cent of the fuel is converted into applicable work. The principal losses of a gasoline engine are: radiation of heat, heat passed off in the exhaust gases, and heat lost by leakages. At the instant explosion takes place in the engine cylinder the temperature at the center of this explosion is estimated to be about 3,000° F. Since cast iron melts at about 2,300°, a great deal of the heat of the explosion must be immediately carried off by radiation through the walls of the cylinder. In order to utilize all the heat left in the gases after the loss by radiation is deducted, the cylinder would have to be so long that the gases could expand to atmospheric pressure. This is a mechanical impossibility. And it has been decided that the most practical length of cylinder is such that the stroke of the piston is about twice as long as the diameter of the engine cylinder. Under these conditions the pressure at release is generally about 40 pounds, and the exhaust gases are still hot enough so that they produce a dull red flame. These two losses are the greatest; and the third loss, that is, the loss past the piston rings, is due to the fact that it is impossible to have a joint between moving parts perfectly tight.

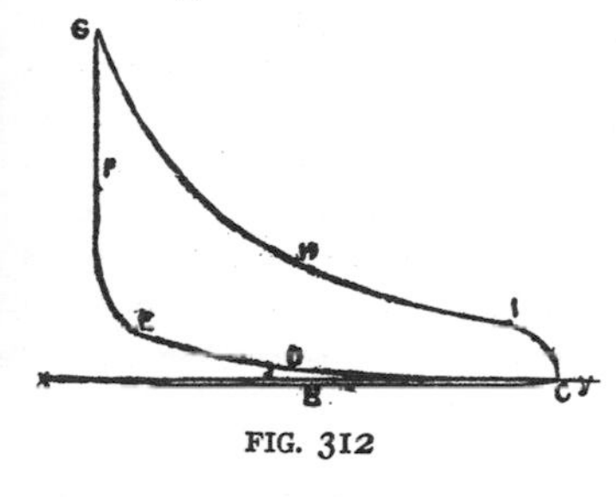

FIG. 312

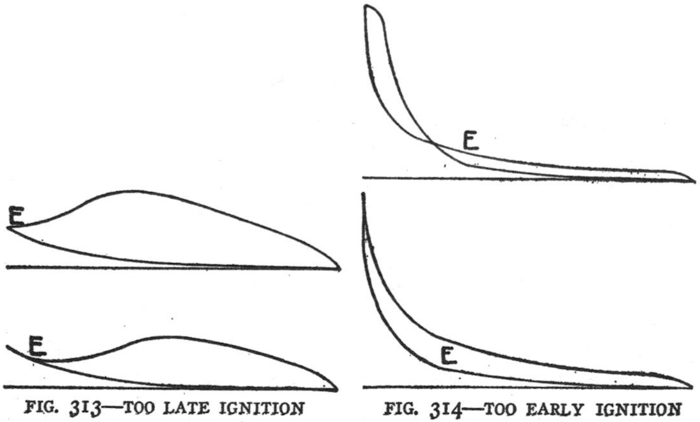

FIG. 313—TOO LATE IGNITION

FIG. 314—TOO EARLY IGNITION

Indicated horse power.—The formula for indicated horse power in the gasoline engine is:

$$H. P. = \frac{PLAN}{33,000},$$

where

$P =$ mean effective pressure,
$L =$ length of stroke in feet,
$A =$ area of piston in square inches,
$N =$ number of explosions a minute.

It will be seen that this formula is the same as that for steam engines, excepting that N represents the number of single explosions a minute in the gasoline engine formula, instead of the number of revolutions a minute, with two impulses for each revolution, as in the steam engine.

Testing.—To make a complete test of a gasoline engine requires a great deal of expensive apparatus. Not only is this apparatus needed, but the one doing the testing must have a very good knowledge of science as far as it pertains to heat and engines. However, a very simple apparatus can be arranged so that any farmer, if he cares to take the trouble, can make a test which will cover all matters as far as he is concerned. The formula for B.H.P. is the same as that given in Chapter I. for steam engines, and the same brake can be used. A speed indicator can be procured for a dollar, and a set of scales or spring balances can be easily secured. It will require two men, who must work simultaneously. Before starting to make the test it will be well to draw up a form about as follows:

Test	1	2	3
Name of engine			
Rated horse power			
Weight of brake on scales, engine still			
Weight on scales, engine loaded			
Net brake load (*G*)			
Length of brake arm in feet (*A*)			
Revolutions per minute (*N*)			
Horse power from test			

Before the engine is started, weigh the brake on the scales with the friction part on wheels ready for the test. Measure the distance from the center of the wheel to the point where the brake rests on the scale. In a rope brake the brake arm is the radius of the wheel plus half the thickness of the rope.

When everything is ready start the engine and gradually draw up the brake. A gasoline engine is running at full load when it misses only one explosion out of about every eight. Tighten the brake until this point is reached, then run the weight out on the scale beam until the point is reached where it balances. Now let one man keep the scales balanced by tightening and loosening the brake; at the same time let the other man take the speed of the engine for one minute. This is all the data needed to determine the brake horse power. It is well, however, to keep the brake on with engine running at full load for at least 15 minutes to determine whether the engine will carry the load.

Care of gasoline engines.—In the care of the engine there are three points of equal importance, namely: cleanliness, water, and oil. To secure the first a well-lighted room is required, one in which the engine alone is placed. Damp, dark cellars should be avoided. As to water and oil, the consideration given depends entirely upon the man in charge. If the engine room is light, the floor clean, waste in a can, tools in a case, and engine bright and clean, it is a certainty that its bearings do not cut for the want of oil, nor its water jacket run dry or freeze up.

A person who does not understand the engine should refrain from tampering with it as long as it runs well. During this time he should be observing and notice the workings of all parts so that in case the engine is not working satisfactorily he can readjust it.

Lubrication.—Lubrication of the gas engine cylinder is very important. A special oil must be used to stand the high temperature met with in gas-engine practice. Any oil containing animal fat will not work at all because when subjected to high temperature it will decompose and be reduced to a charred mass. First-class steam engine cylinder oil will not give good results because it contains certain elements which will carburet like gasoline under high pressure and high temperature. Good engine oil is satisfactory for other parts.

Gasoline engine troubles.—The gasoline engine is often condemned as being unreliable. This may be explained from the fact that unless conditions are just right the gasoline engine will stop or refuse to work at all. This is different in other forms of motors because very often the thing which interferes with its operation comes on gradually and may not be noticed by the man in charge. It has been stated that "there are four things essential to the operation of a gasoline engine," namely: compression, ignition, carburetion or proper mixture, and proper valve action. If these four conditions are obtained, the engine will work or run. If there is failure to obtain any one of them, the engine will refuse to run. Often an engine will stop, and it is difficult to tell which one of the various conditions is wrong. It is necessary to trace the trouble and correct it.

Compression.—It is easy to detect whether or not there is compression by turning the engine over; if a charge of air is caught and compressed, this is an easy matter to determine. Failure to get compression may be due to a valve refusing to seat or to a leak past the valve. It may also be due to a leak past the piston, to broken piston rings, poorly seated rings, or rings gummed with oil. If valves do not seat correctly, it may be due to some gum or scale under one side of the valve. If they leak, it may be due to the fact that the valve seat has become worn owing to excessive heat, in which case they must be reground. If there are broken rings, they must be replaced with new. If poorly seated, new rings must be fitted to the cylinder. If they are gummed up so they will not spring out against the cylinder walls, they may be oiled and loosened with kerosene.

Ignition.—The majority of the gasoline engine troubles may be laid to the ignitor. As stated before, it is often very difficult to pick out the trouble with the ignitor in the case of a battery which has been exhausted

If for any reason the operator thinks the spark fails to pass on the inside of the cylinder, the wire on the insulated terminal should be disconnected and snapped on some bright part of the engine. If there is a spark, it proves that as far as the battery is concerned everything is satisfactory. If there is none, the wire should be thoroughly gone over, the trouble located, and a spark obtained. Perhaps it will be found that a binding screw is loose, or the circuit has been broken at some other point. If the operator gets a spark in the above manner, and then snaps the wire across the insulated binding post, obtaining a spark, there is a connection between the points within the engine, and the ignitor must be removed and cleaned. If by making this test there is no spark, it indicates that there is no circuit between the ignitor points, and the operator should now hold the points together within the engine, by means of the ignitor dog, and snap the wire across the insulated terminal. This time a spark should be obtained, but if not, it indicates that there is insulation between the points, which must be cleaned after the ignitor is removed. Water and carbon will make a circuit between the points, while oil and rust will prevent contact. Any of the above substances between the ignitor points will prevent a spark. The point of ignition varies with the speed of the engine. On a slow-speed engine, one of about 225 revolutions a minute, ignition should take place when the crank is 10° or 15° before center. This angle increases as the speed of the engine increases until in an engine running about 700 revolutions a minute the angle is from 35° to 40°.

To locate ignition troubles is merely a matter of disconnecting certain parts of the circuit and locating the trouble by elimination.

Carburetion.—If for any reason the carburetor refuses to give a proper mixture, the engine will refuse to run. In this case it is necessary for the operator to assure himself that everything else is correct, then clean out the cylinder by turning the engine over several times and beginning as if he were starting the engine for the first time. A too rich mixture is detected by black smoke appearing at the exhaust. A too poor mixture is determined by a snapping sound from the exhaust, indicating that the mixture is slow-burning and is still burning when the exhaust valve opens. The gas engineer determines whether or not his engine is running properly largely from the sound of the exhaust.

Action of valves.—The valves now used on the gasoline engine are all of the poppet type and give a quick opening. An engine will not run if the valves are not properly timed. The inlet valve is operated by suction; however, a little greater efficiency is obtained by having this valve open mechanically, as there is less opportunity for the charge to be throttled during admission. The exhaust valve is always opened mechanically, and should open about 45° before the beginning of the exhaust stroke, closing at the end of the stroke.

Exhaust.—One of the greatest annoyances connected with a gasoline engine is the noise from the ex-

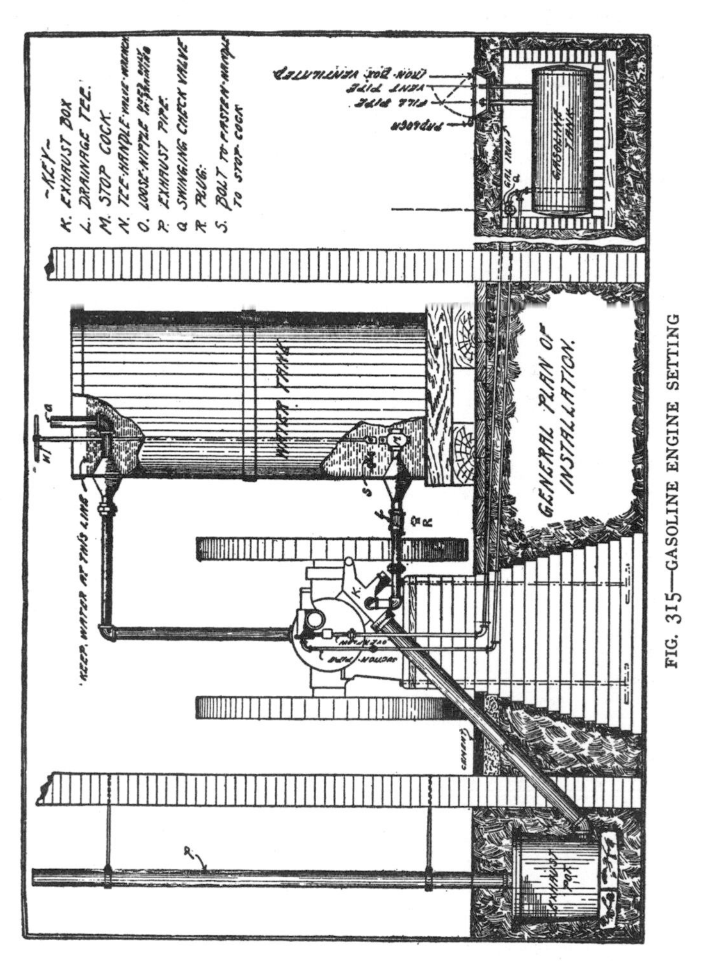

FIG. 315—GASOLINE ENGINE SETTING

haust. All manufacturers send out mufflers or exhaust pots. The latter will not muffle the exhaust appreciably, and the former when muffling effectively generally cause back pressure and consequently loss of power. The most satisfactory method of reducing this noise is to pipe the exhaust into a pit, old well, or smoke stack. The top of the pit or well should be closed, with the exception of three or four openings the size of the exhaust pipe.

Setting.—To insure a smooth-running gasoline engine, the setting is a very important point. Fastening to the ground by means of stakes and skids or to a floor by means of lag screws are makeshift methods. A masonry foundation with well-set anchor bolts is by all means advisable. Well-laid concrete is the best and generally the cheapest. The foundation at the bottom should be about twice the length of the base of the engine and a little more than twice the width. For an engine of 5 to 12 horse power it should be from 3 to 4 feet deep, and for larger sized engines from 5 to 6. The sides should be battered until they are about 8 inches wider at the top than the engine. The jar is to a certain extent broken by having heavy planking between the masonry and the engine. To set and hold the anchor bolts in position, a templet should be made which contains holes corresponding exactly to those in the engine bed. The templet should be made strong and firm. The bolts need a heavy washer or plate at the lower end and should be passed up through a pipe which has an inside diameter of not less than 1 inch. This gives a chance for variation in setting.

Advantages of the gasoline engine as a farm motor.—The gasoline engine has many advantages over the steam engine. In the first place, the farmer as a rule uses power for short intervals. The gasoline engine is always ready to start, and when the run is over there is no fuel in the fire box to be wasted. It does not require an hour's time to get up steam. Not only is there a waste of fire in the fire box, but the steam boiler when under steam contains a large amount of energy, and on cooling down this must all be wasted.

In regard to the matter of safety the gasoline engine has the advantage again. There is practically no danger from explosion with it, for, as was stated, there is not a large amount of energy stored up which may be suddenly released to cause an explosion. Usually the supply tank is placed outside the building, buried in the ground, so the danger from fire is reduced to a minimum. Steam boilers must have an attendant, lest the water get too low and burn the crown sheet, or become too high so water is carried over into the cylinder and knock the cylinder head out. The fire has to be fed continually and the grates cleaned, so that an attendant is needed practically all the time. Such close attention is not needed with gasoline engines.

The gasoline engine is as portable as the steam engine. As to furnishing its own traction, there are several gasoline traction engines on the market, and there is no reason why with the addition of clutches and variable-speed devices the gasoline engine cannot be made as reliable an engine as the steam traction engine. In proof of the fact that it may be made to furnish its own tractive power it is only necessary to refer to the automobile, which is made to work under great variance of speed.

In regard to the cost of power from gasoline and coal, each has advantages under certain conditions. The average consumption of gasoline per horse power per hour should be about 1/6 or 1/7 gallon, with a minimum of 1/10 gallon. The coal consumed per brake horse power per hour is about 8 pounds, with a minimum near 4 pounds as burned under boilers to furnish steam for farm engines. It is possible to figure just what the running expense will be if the cost of the two different kinds of fuel be at hand. Under ordinary conditions and for very small units the gasoline engine will without question be the cheapest. In dairy work, steam direct from the boiler or from the exhaust is used to heat water for washing purposes, and this is a great advantage for the steam plant. However, the jacket water heated with the exhaust of a gasoline engine might be used in the same way.

The steam engine as built for farm use is capable, at the expense of economy, of carrying a very heavy overload. This is extremely advantageous in traction engines in case of emergencies. A 25-horse steam traction engine is often able to develop 60 brake horse power. Gasoline engines are rated very nearly their maximum power, and are not able to carry a large overload.

The troubles with steam engines usually come on gradually, and the attendant is able to observe what is wrong before the engine is stopped. With the gasoline engine, if anything goes wrong the engine stops at once. All conditions must be right in the gasoline engine or it will not run.

The future of the gasoline engine.—Gasoline engines will no doubt be used more and more as time goes on, as they are especially adapted to the farmer's needs. The gasoline engine is a power plant within itself. It can be manufactured in almost any sized unit, and a suitable size can be produced for all manner of farm work from the light work of running grain separators to a motor large enough to run a threshing separator. If gasoline as a fuel becomes too expensive, there is a possibility of a substitution of other liquid fuels in this type of engine.

Engines may be designed to use a heavier kerosene oil, and also alcohol. By the addition of a gas producer, power may be obtained from coal by the use of a gas engine. The internal-combustion engine is the most efficient of all engines; that is, a larger per cent of the heat is converted into mechanical energy than by any other form of prime mover. The efficiency of a steam plant is seldom more than 12 per cent; that of a gasoline engine is not far from 20 per cent. Alcohol works about as well in the gasoline engine as gasoline. The only difficulty to be had is in starting, as alcohol does not carburet as easily as gasoline. As a rough estimate, four gallons of alcohol are equal to three gallons of gasoline.

Alcohol is now manufactured in Germany at about 18 cents a gallon. It is claimed that alcohol can be manufactured as a by-product of sugar factories for as low as 10 cents a gallon. Thus we can feel sure that if gasoline ever becomes so scarce and expensive as to prevent its use upon the farm, we may substitute for it a fuel which may be produced upon the farm itself.

There is a marked advantage in the use of alcohol in that higher compression pressure may be used without pre-ignition. This tends to increase the efficiency of the engine. It is thought that the time will come when every farm will be provided with a power plant in which an engine of the internal-combustion type will be installed.

CHAPTER VI

TRACTION ENGINES

Traction engines.—The steam boiler and the steam engine have been considered separately. If the two should now be combined by means of a steam pipe and placed on skids or trucks they would be termed a portable steam engine. A gasoline engine placed on skids or trucks is known as a portable gasoline engine. Such engines are not self-propelling, but have to be moved by means of animals or some mechanical device. The traction steam engine is the boiler, engine, and propelling device all in one. The traction gasoline engine is the engine and propelling device combined. In other words, the traction engine develops the power by which it moves itself over roads, fields, etc. The action of the traction engine is to convert energy into horizontal motion which has no direct path; that is, the heat from the fuel is transferred from the boiler to the water, then from the water to the steam pipe, and from the steam pipe to the engine, where it is changed from heat energy to mechanical energy. The mechanical energy is then transferred from the engine to the clutch, thence to the drive wheels, which propel the combined unit over its path. The gasoline traction engine is similar to the steam engine in part only and is considered by itself.

ENGINE MOUNTING

In nearly all types of traction engines the engine is mounted upon the boiler, and the boiler is mounted upon the truck. There are now being made some engines which are of the locomotive type, having the engine mounted beneath the boiler. These are known as under-mounted engines.

Boiler mounting.—There are four general methods of mounting the boiler. The most common method is to attach the drive wheels to brackets at the side of the boiler and is known as side mounting. Another common method is to mount the drive wheels on an axle at the

FIG. 316—SIDE-MOUNTED PORTABLE ENGINE

rear end of the boiler and is known as rear mounting. As a rule, the return tubular boilers are mounted on an axle which passes beneath the boiler. This style of mounting is given no special name, but might be called under-mounted boilers. There is now on the market a type of mounting which might be known as frame mounting; that is, there is a frame to which the drive wheels are attached, and it also supports the boiler.

Side mounting.—Fig. 316 shows the method of side mounting a portable engine. This is similar to a great many side-mounted traction engines. Fig. 317

shows a similar side-mounted traction engine. This is done by means of an axle for the drive wheel, which is

FIG. 317—SIDE-MOUNTED TRACTION ENGINE

substantially fixed to a casting. This casting, which is known as the bracket, is then riveted to the side of the fire box. Fig. 318 shows this principle very well excepting that the bracket is strengthened by means of a couple of rods which pass under the fire box and are correspondingly attached to the bracket on the other side. This is a very simple method, but has some disadvantages. The side bracket is attached only to the water leg of the boiler, while the total weight of the engine and boiler is thrown

FIG. 318—BRACKET FOR SIDE-MOUNTED ENGINE

upon it. It is obvious that this puts undue strain upon the boiler shell at a point where it is the weakest. The weight of the boiler and the engine is thrown upon these brackets and in such a manner that it has a tendency to throw the inside of the axle down and the outside up. This will tend to throw the tops of the drive wheels together and the bottoms apart. The weight is also thrown upon these axles so that that part of the hub of the fly-wheel next to the engine will wear faster than the middle, and as a result the wheels will tend to become wobbly in action and wear the teeth of the transmission gearing unevenly. A truss bar similar to that of Fig. 318 removes a great deal of the strain from the water leg, and

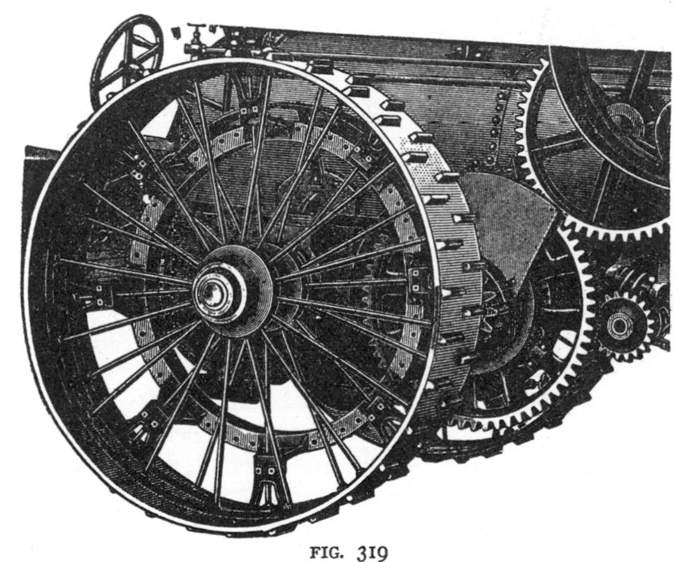

FIG. 319

also tends to hold the axles in line with each other, and thus keep the drive wheels more nearly vertical. Another method of side mounting an engine is shown in

Fig. 319. By inspection it will be noticed that this style of mounting is similar to that of Fig. 317, but in addition to this there is a heavy curved axle which passes from the bracket down beneath the fire box and up to the bracket on the opposite side. Although this style of mounting is considered superior to the one previously described, in order to prevent springing of the axle and the consequent wobbling of the wheels it will be necessary to make the axle too heavy for practice. Although the bad effects of the strain on the boiler are practically all removed by passing the axle beneath the fire box, the effect of the wearing of the boxings in the hubs is still uncared for. This allows the wheels to travel out of a vertical plane and wear the gearing irregularly. Fig. 320 shows an end view of this style of mounting, with the addition of springs. These springs are a benefit to a traction engine in that they take the jar off the parts as the engine travels over rough roads or pavements.

FIG. 320—SIDE-MOUNTED ENGINE WITH SPRINGS AND TRUSS BAR

Rear mounting.—Rear mounting, as a rule, is not as simply done as side mounting. However, it has some advantages over the other. Fig. 321 shows one type of rear mounting which has its merits. The brackets which support the boiler and the engines are attached to the corners of the water leg, thus removing the strain from a weak point to one which is stronger. By having the engine rear-mounted the axle upon which the drive wheels travel is allowed to revolve in the bearings instead of the wheels revolving upon the axle. By having the axle revolve in this manner the wear is all in a straight line and on the top of the boxing, hence there is no reason for the wheels to become wobbly and cut the transmission gear-

FIG. 321—REAR-MOUNTED ENGINE

ing unevenly. By referring to Fig. 322 it will be seen that the use of springs becomes impossible on a rear-mounted engine as shown in Fig. 321. Assuming that there are springs in this type of mounting, and that the springs are so adjusted that when a jar comes upon the engine the teeth will mesh as shown in Fig. 322, then if there were no jar upon the engine and the springs were carrying it in its normal position, the gears *A* and *B* would not mesh, or else they would mesh just enough so that the teeth would catch and strip. If a spring could be placed so the combination of gears *A*, *B*, and *C* would

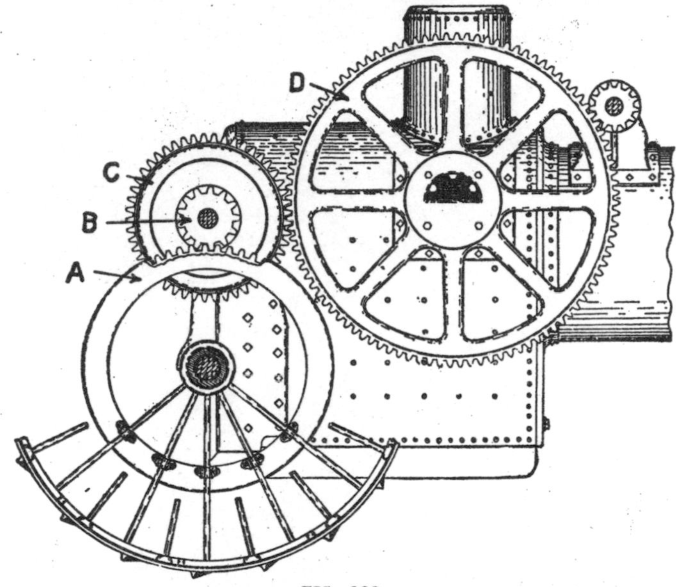

FIG. 322

rise and fall together in a circle whose radius is equal to the sum of the radii of the wheels *C* and *D*, it would be as effective and the wheels *C* and *B* would mesh. Fig. 323 shows the type of mounting which has this desired effect, but it has the additional complication of radius and cross links. As the springs respond to the jars of rough roads, these links keep the gear wheels the proper distance apart, so that they are always in proper mesh.

FIG. 323—REAR-MOUNTED ENGINE WITH RADIUS AND CROSS LINKS

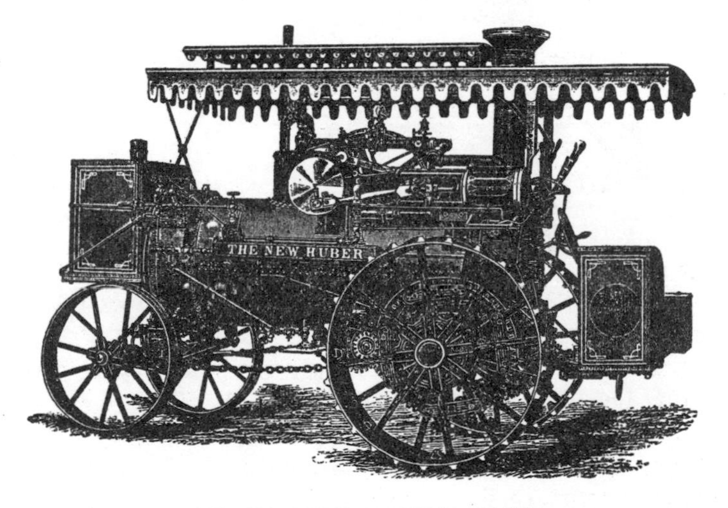

FIG. 324—UNDER-MOUNTED ENGINE

Under-mounted boilers.—Fig. 324 shows a type of mounting where the axle is straight and fastened directly beneath the boiler. By inspecting Fig. 325, this method of mounting will be more clearly understood. *A* is the

FIG. 325

FIG. 326—FRAME-MOUNTED ENGINE

main axle upon which the drive wheels operate. Although made of a square bar, it is round at the bearing *B*, and revolves in it. Although the brackets for this type of mounting are attached to the boiler, the boiler itself, being round, is probably strong enough so that the excessive strain will cause very little trouble. This type of mounting very seldom contains springs.

Frame mounting.—To remove as much of the strain as possible from the boiler, some engines are now coming upon the market with a frame which supports engine, transmission gears, and boiler. Or else the frame supports the boiler and the boiler supports the engine.

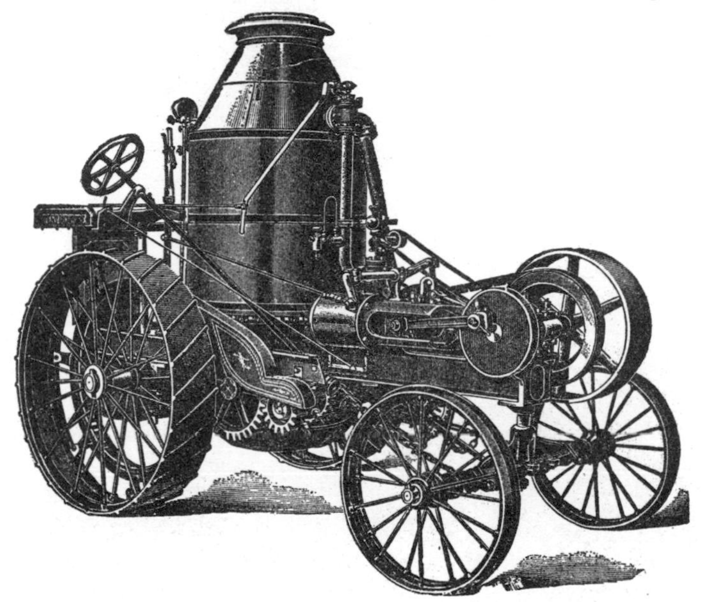

FIG. 327—FRAME-MOUNTED VERTICAL TRACTION ENGINE

Fig. 326 shows the frame for this type of engine with the boiler and transmission gears removed. Fig. 327 shows a vertical traction engine and boiler complete. For a certain class of work there is a call for a style of frame mounting such as is seen in Figs. 328 and 329. In this style of mounting all the strain is thrown upon the frame, allowing the boilers to be freely suspended as shown.

FIG. 328—FRAME-MOUNTED ENGINE OF THE LOCOMOTIVE TYPE

FIG. 329—LOCOMOTIVE TYPE OF TRACTION ENGINE WITH STEAM-OPERATED PLOW

Engine mounting.—Where the engine is not mounted upon the frame as shown in Figs. 326, 327, 328, and 329, it is mounted upon the boiler. This is not considered the best method. However, it is commendable for its simplicity and possibly counterbalances the evil effects of the extra strain upon the boiler. Fig. 330 illus-

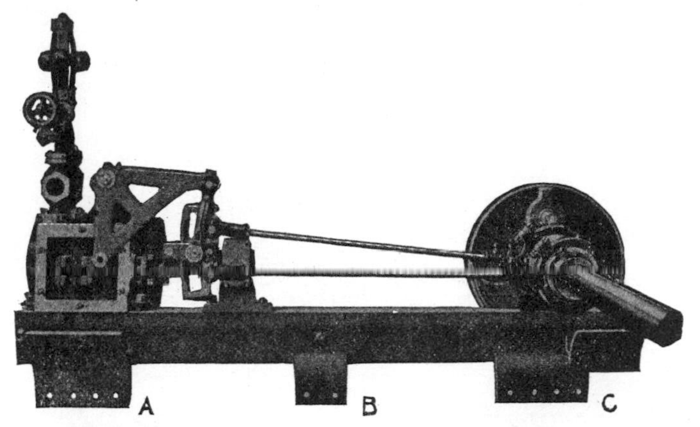

FIG. 330—ILLUSTRATING METHOD OF MOUNTING ENGINE ON BOILER

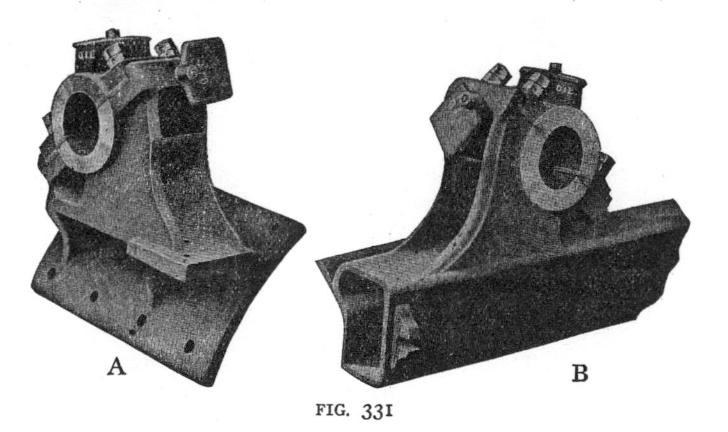

FIG. 331

trates the method which most engine builders utilize in attaching their engines to boilers. The brackets *A*, *B*, *C* are riveted directly to the boiler shell.

Fig. 331 shows the main bearing *A*, which is a part of the frame, also the bearing *B*, which is commonly known as the pillow block bearing. These bearings are both riveted to the boiler.

Clutch.—When the separator is being driven by the engine the traction part must not move. Consequently, there must be some method for throwing the power from the drive wheel which drives the pulley to the transmission gearing that runs the traction part of the engine. The device for transferring the power is a clutch generally located on the engine shaft. It acts upon the belt wheel of the engine. Fig. 332 shows a sim-

FIG. 332—CLUTCH

ple clutch in parts. *A* is the belt wheel upon which travels the belt that drives the separator. It is fixed to the engine shaft so that whenever the engine moves this wheel moves also. The clutch blocks and arms are seen at *D*, and the pinion is engaged with the transmission gearing at *C*. This part of the mechanism is not fixed to the shaft, and revolves with it only when the clutch is locked. In other words, when the clutch locks, the blocks all are forced out against the rim of the belt wheel tight enough so they stick to it and the whole mechanism revolves with the wheel. The clutch is a very important part of the traction engine and requires very careful adjustment and care. Since the blocks *DDD* are continually wearing off, the arms *EE* have to be constantly adjusted. They should be so carefully adjusted that when thrown in, the clutch will lock and hold itself in position. They should also be adjusted so there will be no slip between the clutch blocks and the clutch shoe. Fig. 333 shows another type of clutch, which has a metal clutch block instead of wood.

FIG. 333—CLUTCH WITH METAL BLOCKS

Transmission gearing.—The steam engine for traction engine work generally has a speed of 200 to 225 revolutions a minute. If the drive wheels were connected directly to the engine shaft such a speed would drive the outfit over the ground nearly as fast as a locomotive travels. This is something that could not be conceived of on country roads, hence the speed has to be reduced to one which is permissible. For this purpose a chain of gears such as is shown in Fig. 334 is utilized. Not only are these gears used to reduce the speed of rotation from that of the drive wheel to that of the engine, but since the engine is generally located some distance from the traction wheel shaft, these gears conduct the power from the engine shaft to the shaft of the traction wheel. The intermediate gears are generally attached to the boiler by means of brackets as shown in Fig. 322. If the engine were always to travel straight ahead or straight backward the matter of transmission gearing would be very simple, but since it has to turn and often on a very small circle one wheel is compelled to travel faster than the other; consequently they cannot be both attached rigidly to the same shaft.

FIG. 334—GEARS CONNECTING ENGINE WITH TRACTION WHEELS

If one wheel were attached to the shaft and the other were allowed to go free then one wheel would do all of the traction work. This would not do, since the engine would have only half of the tractive power and for road work it is necessary that every pound possible of tractive pull be developed. To arrange the drive wheels of a traction engine so that both will pull when the engine is traveling in a straight line and also so they will travel in a curve without slipping, a compensating gear is inserted in the chain of transmission.

Compensating gears.—Fig. 335 shows a simple, very effective compensating gear. The large pinion *A* carries the small pinions *CCC*. The shaft *F* is connected to the flywheel on the opposite side of the engine by means of a small pinion. The pinion *G* is connected to the other main gear. The power is transmitted from the engine shaft to the pinion *A*. As pinion *A* revolves in the direction of the arrow, pinions *CCC* will be driven, and they in turn will propel the drive wheels. But if the drive wheel attached to pinion *G* happens to travel faster than that attached to shaft *F* the pinion *C* will revolve and still the pinion *A* will propel the gearing. Often there are some very severe jerks on the transmission gearing of an engine and some companies are now inserting in their compensating gears a set of springs which take this jar off the gearing.

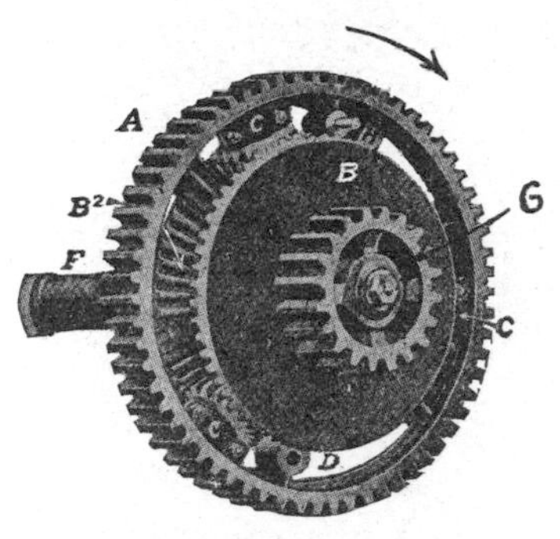

FIG. 335—COMPENSATING GEARS

Traction.—Any traction engine has power enough to propel itself over the road and through the fields provided the drive wheels do not slip. Consequently the matter of the wheels adhering to the ground is an important part. Where the road surface is firm there is no difficulty; but in a soft field great trouble is experienced due to the fact that the lugs of the drive wheels tear up the earth and allow the drive wheels to move without moving the engine. It is a common belief that the drive wheel which has the sharpest lug is the one which will adhere to the ground the best. In nearly all cases this is not true, since the lug which is sharp is very apt to cut through the earth, while one which is dull or round and does not have such penetrating effect will pack the earth down and thus make more resistance for itself while passing through the earth. Nearly every engine builder has a style of lug of his own. Fig. 338 shows a new style of traction wheel which seems to be giving very good results. The more weight that can be put on to the drive wheels of an engine the better it will adhere to the ground, providing the surface is firm enough to support the load. This makes the matter of location of the main axles upon the boiler an important factor. When the boiler is rear-mounted it is obvious that more of the weight is thrown upon the front wheels, which act as a guide, than when the boiler is side-mounted. Hence one would be led to believe that the side-mounted traction engine will have better tractive power than the rear-mounted. It is also indicative of better tractive power when the pivot of the front axle is as far ahead as possible. For this reason some builders are now attaching a frame to the boiler and crowding the front trucks ahead. Fig. 336 is an illustration of this type of mounting.

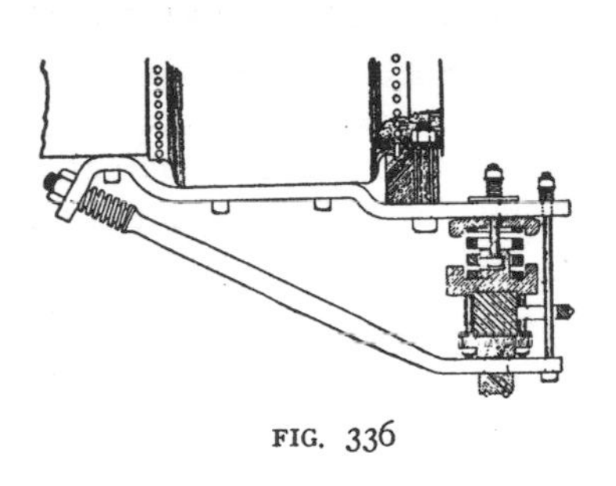

FIG. 336

Width of tires.—Where traction engines such as are used for harvesting and threshing grain simultaneously are used for plow work or in the field an exceptionally wide tire is required. If an engine is to be used for this work exclusively the wheels are made with the proper width of tires at the factory. But where an engine is to be used for job threshing a part of the time and for plowing a part of the time the wheels should be made so an extra width of tire can be attached to support the engine for plowing.

Road rollers.—For road rolling purposes traction engines as a rule, especially the gearing and bearings, are made much heavier. The tires are wider, and the front truck instead of being made of two wheels is made into one broad wheel.

HANDLING A TRACTION ENGINE

Moving an engine.—When moving an engine it is best to carry more water than when doing stationary work. This is especially true in hilly fields or hilly roads. The gauge glass and water cocks should be carefully watched. The steam pressure should be maintained near the blow-off point. Upon approaching a hill judgment should be exercised in regard to the fire and amount of water and pressure. As much water should be carried as is permissible without priming. If possible there should be sufficient fire when starting up a hill to carry the engine to the top. Judgment should also be exercised in regard to the speed. Taking an engine up a hill too fast is apt to cause priming. Also there is danger of reducing the steam pressure so that a stop will have to be made to raise it. When the summit of the hill has been reached, the fire can be started up, more water put in the boiler, and the engine allowed to travel faster. As much and probably more care should be exercised in descending a hill than in ascending. If possible the engine should be taken from the top of the hill to the foot without a stop. If this is not possible turn the engine around so that it sits as near level as possible while the stop is being made. Every engineer knows the danger of having the front end of a fire box boiler the lowest. If the engine is inclined to run too fast in going down a hill the reverse should be thrown. If then it still travels too fast, while the engine is still in the reverse, open the throttle and let in a little steam.

Guiding an engine.—Traction engines are guided by means of the hand wheel, which operates through a worm gear. This in turn acts upon chains which are attached to the ends of the front axle. Turning the hand wheel to the right will turn the engine in that direction, while in turning the hand wheel to the left the engine will turn to the left. Do not turn the steering wheels too often or too far. Watch the front axle and act accordingly. It is much easier to steer an engine when moving than when standing. If possible always move the engine a trifle when steering. The steering chains should be moderately tight; if they are too tight they will cause undue friction, while if they are too loose the engine cannot be guided steadily.

Mud holes.—The best way to get out of a mud hole is not to get into it. An engineer should go out of his way a considerable distance rather than to take his engine into a mud hole. When an engine is once in a mud hole and the drive wheels commence to slip without propelling, the engine should be shut down at once. When the drive wheels are run in the mud without moving the engine they soon dig up a hole out of which it is very hard to raise the engine. When drive wheels commence to slip, straw, boards, rails, posts, or anything at hand should be put under them so they may get a grip. In getting out of a mud hole do not start the engine quickly, but very slowly. If the wheels will stick at all they will gradually move the engine by starting it slowly, while if starting it quickly the grip of the wheels gives away before momentum can be put into the engine. If stuck in a mud hole always uncouple the separator or whatever load the engine is hauling, move the engine out, then by means of a rope or chain pull the separator across. If the engine is stuck in a soft place like a plowed field often the hitching of a team in front will take it out.

Bridges.—Before crossing a bridge or culvert the engineer should make inspection to see if it will carry the weight of his engine and the separator. If there be any doubt and it is impossible to move the engine around the bridge heavy planks should be placed across it to distribute the load. Always move slowly while crossing a bridge. If the engine has once broken through it can sometimes be removed by winding a rope around the belt wheel several times, then setting the friction clutch and hitching a team upon the rope. As the rope gradually unwinds, it will move the engine by means of the transmission gearing.

Gutters.—In road work often one drive wheel of an engine will strike a soft place in the gutter. Owing to the principle of the compensating gear this wheel will then slip in the mud and revolve while the other wheel will remain stationary and the engine not move. In a case like this the compensating gear should be locked and both wheels be made to revolve together. The wheel which is on the solid ground will move the engine out of the hole. To lock the compensating gear there is generally some scheme, as in Fig. 335, whereby a pin can be inserted in the pinion *A* and lock the pinion *D* by means of the projection *H*.

Reversing the engine on the road.—When it is desired to reverse a traction engine moving on the road the throttle valve should be closed, the engine reversed, then the throttle opened. Traction engines are usually made strong enough so they will stand the strain of being reversed without closing the throttle. This, however, is hard on the bearings, and the engineer should always close the throttle before reversing the engine, especially if the engine is running at full speed.

Setting an engine.—A new engineer will experience some difficulty in setting an engine so it is properly lined with the separator. On a still day the belt wheel of the engine should be in line with the separator. This is also true when the wind is blowing in line with the engine and the separator. But if the wind is at an angle allowance will have to be made for the amount which it will carry the belt to one side. Often the engine will have to be set a few feet out of line with the separator and toward the wind. If the engine has been set when there is no wind and enough wind comes up to throw the belt over, it is not necessary to stop the engine and move, but a jack screw can be set against the end of the front axle and the engine worked over toward the wind. Also the front end of the separator should be crowded in a similar manner until the belt runs in the proper position on the pulley. The friction clutch should always be used in backing the engine into the belt.

Gasoline traction engines. — Since the gasoline traction engine requires no boiler, the engine with its necessary accessories, such as water tanks, gasoline tanks and battery boxes, is mounted upon a frame. Consequently the mounting of a gasoline engine is more simple than that of a steam engine. However, it has a disadvantage which the steam engine does not have; that is, the engine itself cannot have its direction of rotation reversed without a great deal of trouble, consequently there has to be connected into the transmission gear a reversing gear. The simplest of the reversing gears for gasoline engines now on the market is a system of friction pulleys, such that when the engine is in one position on the frame the traction wheels will move forward. When it is in another position another set of wheels is connected in and the traction wheels will move backward. It will be noticed from this that the engine, which generally weighs 2,000 or 3,000 pounds, has to be slid backward or forward on the mounting frame. Fig. 337 shows a type of engine which reverses as above

FIG. 337—GASOLINE TRACTION ENGINE WITH FRICTION GEARING

described. This engine is operated by means of a set of friction wheels, instead of a set of gearing as steam traction engines are run. Fig. 338 illustrates an engine which utilizes pinions for its transmission gearing similar to a steam traction engine.

Rating.—Gasoline traction engines are all rated upon

the horse power they will develop at the brake. Consequently when one speaks of a 15 H.P. gasoline engine he refers to an engine which will develop only about the same horse power which a commercially rated 7 H.P. steam engine will develop. For this reason when comparing the powers of the two engines it is always well at least to double the size of the gasoline engine to do the work which a commercially rated steam traction engine has been doing.

FIG. 338—TRACTION ENGINE WHICH REVERSES IN THE CLUTCH

Regulation of speed.—A gasoline traction engine operated by means of friction gearing, as illustrated in Fig. 337, can have any speed required of it at the expense of slippage between the gears. But a positively driven traction engine must have other methods of changing the speed. These methods generally amount to changing the point of ignition in the engine in order to reduce the power at low speed, or else shifting the power from one set of gears to another. Generally in an engine where the power is shifted there are only two speeds, a high and a low.

On the road.—About the same caution should be exercised in handling a gasoline traction engine through soft and muddy places and over bridges as in handling a steam engine. But there is practically no caution to be taken in climbing hills other than that taken on level ground. Upon descending a hill a strong and effective brake should always be at the control of the operator.

Traction.—As a rule gasoline traction engines are much lighter than steam traction engines. Consequently their tractive power is correspondingly less. And for heavy traction work the size of the engine must be increased in order to add to the tractive power.

CHAPTER VII

ELECTRICAL MACHINERY

Natural magnets.—The name magnet was given by the ancients to a brown-colored stone which had the property of attracting certain metals. Later the Chinese found that when free to move this stone always pointed in one direction, and they named it loadstone (meaning to lead). The commercial name for it is magnetite (Fe_3O_4). This mineral is found in such quantities in several localities that it is a valuable ore for producing iron.

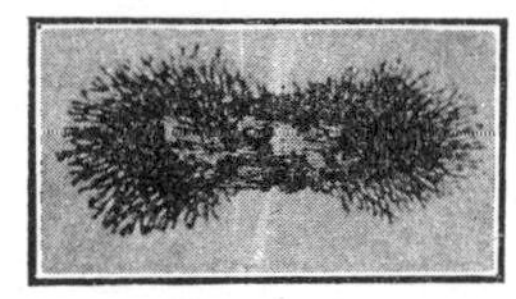
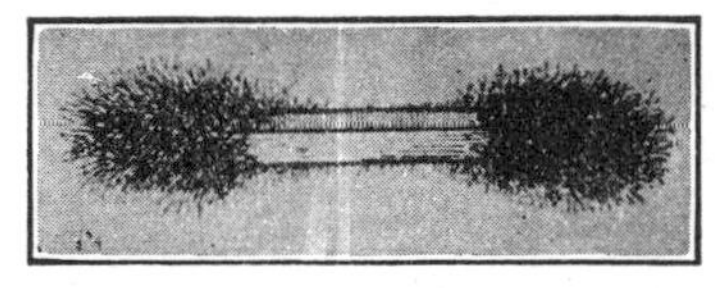

FIG. 339—NATURAL AND ARTIFICIAL MAGNETS ATTRACTING IRON FILINGS

Artificial magnets. — The ancients learned by stroking pieces of steel with natural magnets that the steel would become magnetized. Magnets produced in this manner are known as artificial. They are now made by stroking bars of steel with another magnet or an electromagnet, which will be described later.

Poles.—If a magnet is sprinkled with tacks or iron filings, it will be noticed that the filings attach themselves to the ends of the magnet but not to the middle of it. The name poles has been given to these places where the filings adhere. A suspended magnet will swing so that one of its poles points toward the north. This pole is then known as the + or north-seeking pole, or simply the north pole (N), and the other end is known as the —

or south pole (S). The mariners' and the engineers' compasses work upon the same principle.

Magnetic lines of force.—Again, if a sheet of paper be placed over a magnet and some filings then dropped upon the paper, and if the paper is slightly jarred, the filings will assume the position shown in Fig. 340. From this it is gathered that the magnet has lines of force and that these lines are of the form indicated in Fig. 341. For convenience it is assumed that the lines of force leave the magnet at the N pole and enter at the S pole.

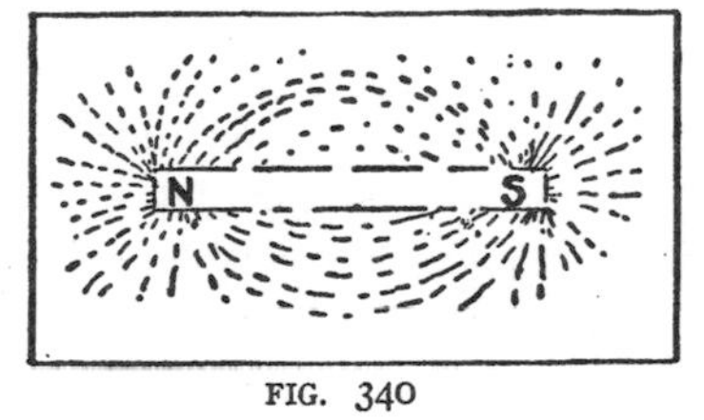

FIG. 340

Laws of magnets.—If the north and the south poles of two magnets are determined and marked it will be noticed that when one of the magnets is suspended so it is free to move in any direction and the north pole of the other is brought close to the south pole of the suspended one, these two ends attract each other. If, on the other hand, the N ends be brought together it will be noticed that they repel. Hence the general law of magnets is deduced: **Like poles repel and unlike poles attract.**

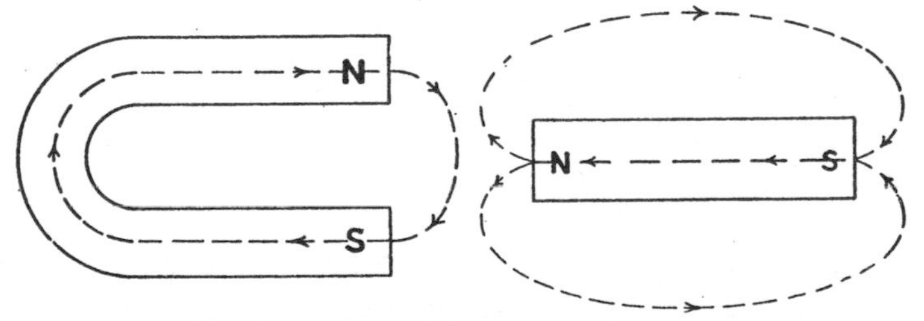

FIG. 341—DIRECTIONS OF LINES OF FORCE

The force of this attraction is found to vary inversely as the square of the distance, i.e., increasing the distance between the poles two times reduces the force acting between them $2 \times 2 = 4$ times. In other words, the force is one-fourth as strong.

Magnetic materials.—Steel and iron are the only common substances which show magnetic properties to any appreciable degree.

STATIC ELECTRICITY

Static electricity.—If a hard rubber rod be rubbed with flannel and then brought close to a suspended pith ball the ball will jump toward the rod. By rubbing the rod has been electrified and the action of the charge is to attract the ball. This charge of electricity is not within the rod but is on the surface and is known as stationary or static electricity. Another example of this is rubbing a glass rod with silk.

Laws of electrical attraction and repulsion.—If a rubber and a glass rod be excited and suspended as shown

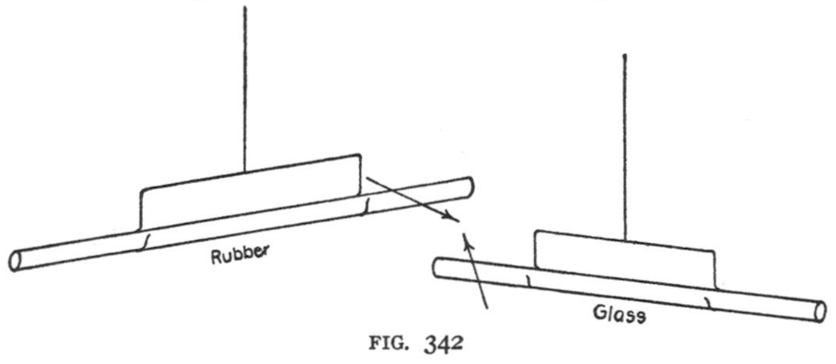

FIG. 342

in Fig. 342 and brought close together it will be noticed that they attract each other, but if two rubber rods be suspended in the same way and brought together, they will repel each other. Hence the following law is advanced: **Electrical charges of a like kind repel each other and those of an unlike kind attract.**

Density of charge varies with form of surface.—

Since all of the little particles of a charged substance, because of their mutual repulsion, tend to get as far away from each other as possible, the density of a charge is very much greater on the ends of an oblong body than in the middle. If the ends be drawn to a point the charge will become so intense that the point cannot hold it all and some of it will be given off to the air.

Lightning and lightning rod.—In 1752 Franklin with his famous kite and key learned that there is electricity in the clouds. He also showed that lightning is only a huge electric spark and that by means of points like lightning rods these mammoth sparks may be dissipated into the earth. As the cloud which is charged with electricity approaches it induces an opposite charge in the points and the charge is then quietly conducted away, while if the points are not there the electric charge will assume such a volume that when the cloud does give it up it will strike the building in such a great bolt that damage is done. From this it will be seen that lightning rods do not protect the building by conducting the whole charge of the stroke away at once, but by diffusing and thus preventing the charge collecting in large quantities.

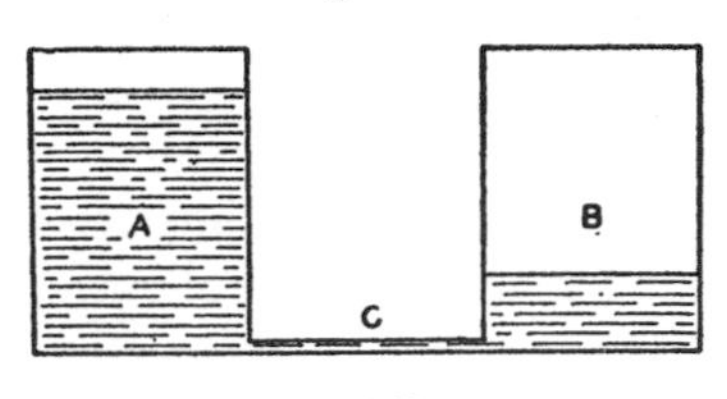

FIG. 343

Potential difference (P.D.).—If water is placed in a tank *A*, Fig. 343, it will run through the pipe *C* into tank *B*. We attribute the running of the water from tank *A* to tank *B* to the difference in pressure between the two tanks. In exactly the same way will a positive charge of electricity flow from one body to another. Thus, just as water tends to flow from higher pressure to lower, does electricity of a higher potential flow to a lower. Moreover, if the tank *A* is not continuously supplied with water the tank *B* will soon be filled to an equal level; likewise if current is not supplied to the body having the greater potential, the potential will become the same in the two bodies.

Volt or unit of potential difference.—To measure the amount of work required to transfer a charge from one body of a high potential to one of a low potential there must be a unit. This unit is called the **volt** in honor of the great physicist Volta. It is roughly equal to the P.D. between one of Volta's cells and the earth.

CURRENT OR GALVANIC ELECTRICITY

Current electricity.—Electricity is an invisible agent and is detected only by its effects or manifestations. Current electricity is generally detected by its magnetic effects. That is, near all currents of electricity there are indications of magnetism, while in stationary or static electricity there are none.

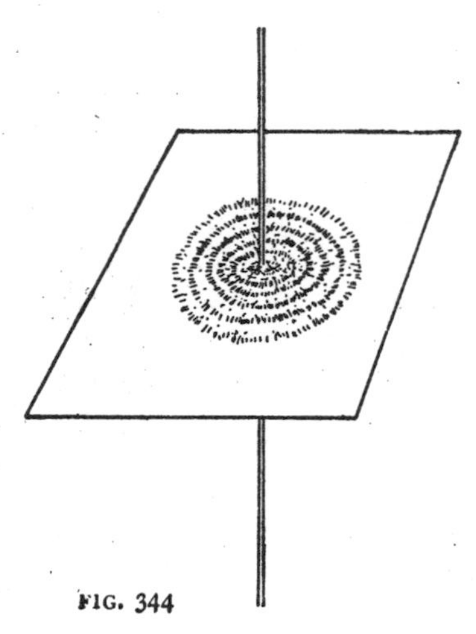

FIG. 344

Shape of magnetic field about a current.—If a wire carrying a heavy current of electricity be run through a cardboard and filings be sprinkled upon the board they will form themselves into concentric rings about the wire (Fig. 344). A compass placed in this field and at several positions will show that the lines of force are all in one direction. Reverse the current and the needle will also reverse. This shows that there is a direct relation between the direction of the current in the wire and the direction of the magnetic lines which encircle it.

Right-hand rule.—Ampère devised a rule in which the right hand is used as a means to indicate this relation in all cases. Let the right hand grasp the wire (Fig. 345) so that the thumb points in the direction in which the current is flowing and the fingers will then point in the direction of the magnetic lines of force. Ampère being the investigator who made quantitative measurements of current electricity, the unit of measurement was named **ampere** in his honor. Owing to the peculiarity of electricity it cannot be measured in pints and gills as liquids but can be measured by the chemical effect it will produce, i.e., one ampere will deposit in one second 0.0003286 gram of copper in a copper voltmeter.

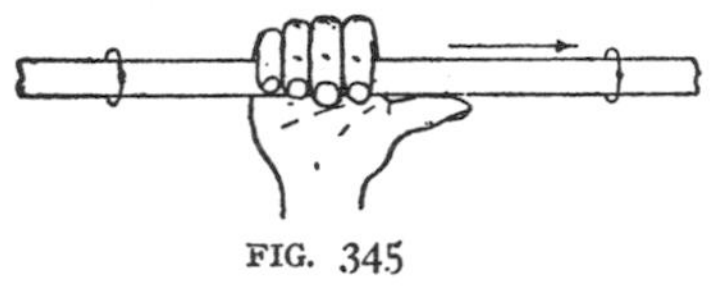

FIG. 345

The ammeter is an instrument used for the measuring of amperes. Commercial ammeters do not measure them by means of chemical deposits, but by means of a needle enclosed in an electrical coil in such a

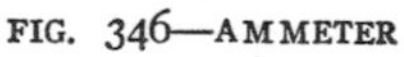

FIG. 346—AMMETER

FIG. 347—VOLTMETER

manner that as the current varies the magnetic force of the coil will vary, and cause a deflection of the needle.

Voltmeter.—To measure the electrical pressure or potential difference requires an instrument similar to the ammeter excepting that instead of having a few coils of wire it often has several thousand coils of very fine wire. Only a very small amount of current will pass through these numerous coils.

Electromotive force.—The total electrical pressure which an electrical generator is able to exert is called its electromotive force, commonly abbreviated to E.M.F.

Electrical power.—The unit of electrical power is a unit of electrical work performed in a unit of time and is called a **watt.**

The product of volts into amperes gives watts, i.e., volts $\times$ amperes $=$ watts.

Example.—An incandescent lamp is fed by a current having a voltage of 220 and requires 0.3 ampere of current. The electrical power consumed is then

$$V \times A = W,$$
$$220 \times 0.3 = 66.0 \text{ watts.}$$

Kilowatt.—The watt is such a small quantity that it has become the custom to use a larger unit known as the kilowatt.

$$1 \text{ kilowatt} = 1{,}000 \text{ watts,}$$

or,

$$1 \text{ watt} = 1/1{,}000 \text{ kilowatt.}$$

Horse power.—By experiment it has been found that 7375 foot pound per second = 1 watt.

Now, since 550 foot pounds a second is the equivalent of one mechanical horse power, an equivalent rate of electrical working would be:

$$\frac{550}{.7375} = 746 \text{ watts} = \text{one electrical horse power.}$$

Resistance.—If two pipes of the same diameter but different lengths lead from a tank of water, the water will flow very much faster from the short pipe than from the long one. From this we learn that the pressure decreases as the water passes through the pipes and the longer the pipe the more it falls. The friction between the water and the inside of the pipe retards the flow and is known as resistance. Electricity flowing over a wire is an analogous case. The current meets with resistance in the wire and there is a fall in potential.

Comparative resistance.—To measure comparative resistance, silver is the unit of comparison, it having the lowest resistance of any substance.

Specific resistance of some metals:

Silver, 1.00;
Copper, 1.13;
Aluminum, 2.00;
Soft iron, 7.40;
Hard steel, 21.00;
Mercury, 62.70.

Laws of resistance.—As the lengths of wire increase the resistance increases and as the diameter increases the resistance decreases. Hence the following law is deduced: That the resistance of conductors of the same materials varies in direct proportion to length and inversely to the area of the cross-sections.

The resistance of iron increases with rising temperature, likewise with nearly all metals, while the resistance of carbon and liquids decreases as the temperature increases.

Unit of resistance.—A conductor maintaining a P. D. of one volt between its terminals and carrying a current of one ampere is said to have a resistance of one ohm. The **ohm** is the unit of resistance and is named in honor of George Ohm, the great German physicist.

Ohm's law.—The current existing in a circuit is always

directly proportional to the E.M.F. in the circuit and inversely proportional to the resistance.

Hence if

$$C = \text{current},$$
$$E = \text{E.M.F.},$$
$$R = \text{resistance},$$
$$C = \frac{E}{R}, \text{ or current} = \frac{\text{E.M.F.}}{\text{Resistance}}.$$

Likewise,

$$\text{Amperes} = \frac{\text{Volts}}{\text{Ohms}}.$$

Rheostats.—The common method for controlling the current required for various electrical purposes is either to insert or to remove resistance. By Ohm's law

$$C = \frac{E}{R}. \qquad (A)$$

If E is kept constant and R is varied, C will also be varied but with an inverse ratio. Any instrument which will change the resistance in a circuit without breaking it is known as a **rheostat.** A rheostat can be constructed

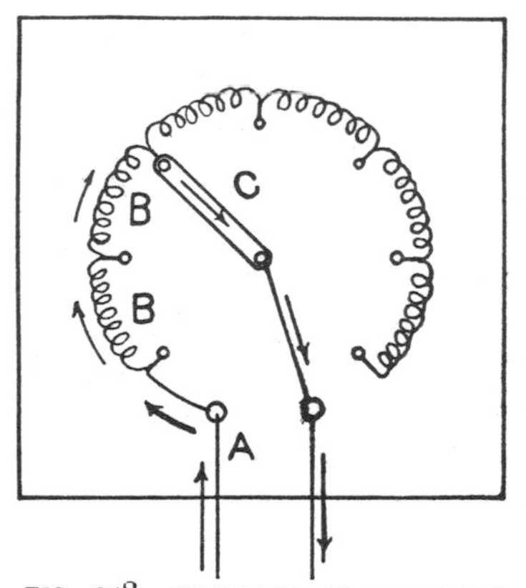

FIG. 348—PRINCIPLE OF RHEOSTAT

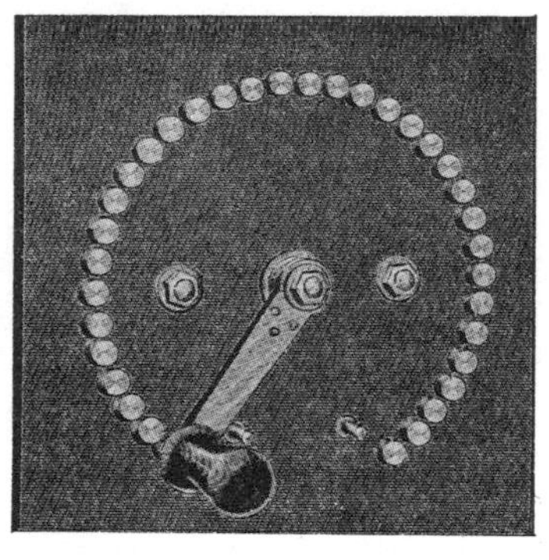

FIG. 349—COMMERCIAL RHEOSTAT

of various substances: coils of iron wire, iron plates or strips, carbon, columns of liquids, etc. Fig. 348 illustrates a commercial rheostat. The current enters at *A*, passes through the resistance B, which can be increased or decreased as the metallic arm C is moved from point to point, and out through the arm C and pivot D. The rheostat absorbs energy and throws it off as heat instead of doing useful work with it.

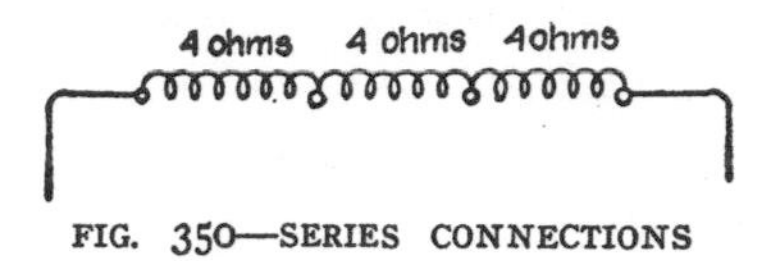

FIG. 350—SERIES CONNECTIONS

Series connections.—When lines are connected up as in Fig. 350, so that the same current flows through each one of them in succession, they are said to be connected in series. In this case the total resistance is the sum of the several resistances.

$$4 + 4 + 4 = 12.$$

Parallel connection. — If instead of connecting these lines up as in Fig. 350 they be connected as in Fig. 351 they will be in parallel and the total resistance will be only one-third of the resistance of one of them. This is obvious, for in this connection there is three times as much cross-section of wire carrying the current as in the previous case, and by formula (A) the resistance varies inversely with the sectional area.

Shunts.—One line connected in parallel with another is said to be a shunt connection to the other. In Fig. 351A, S is shunted across the resistance R. If R has a greater resistance than S it will carry less of the current, since the currents carried are inversely proportional to the resistance. Hence if R has a resistance of 5 ohms and S a resistance of 1 ohm, R will carry one-fifth as much current as S or one-sixth of the total current.

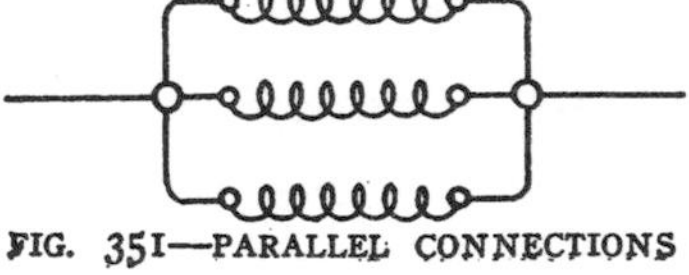

FIG. 351—PARALLEL CONNECTIONS

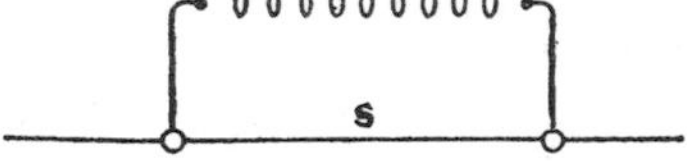

R
S

FIG. 351A—SHUNT

Cells.—If a strip of copper be connected to one end of a strip of zinc and the free ends of the two metals be immersed in dilute sulphuric acid (Fig. 352) a current of electricity will manifest itself in the wire. If the circuit is broken and the plates carefully watched, bubbles will be seen to collect on the zinc plate and none on the copper. As soon, however, as the circuit is completed again a current will be noticed, also a great number of bubbles will appear about the copper plate. These last bubbles are bubbles of hydrogen and always appear when a current is being produced. The bubbles which form about the zinc are also of hydrogen, but they are caused by the zinc being impure and by a current starting up between these particles of impurities and the particles of zinc. This action is detrimental to the cell and should be stopped by covering the zinc with mercury. By permitting the current of this cell to run for some time it will be noticed that the zinc is being gradually eaten away, and that the copper plate does not change. From this it is learned that when the current of a simple cell is formed the zinc is eaten away and hydrogen collects on the copper. The current passes out from the copper plate and in on the zinc. In other words, the copper plate is the positive terminal and the zinc is the negative.

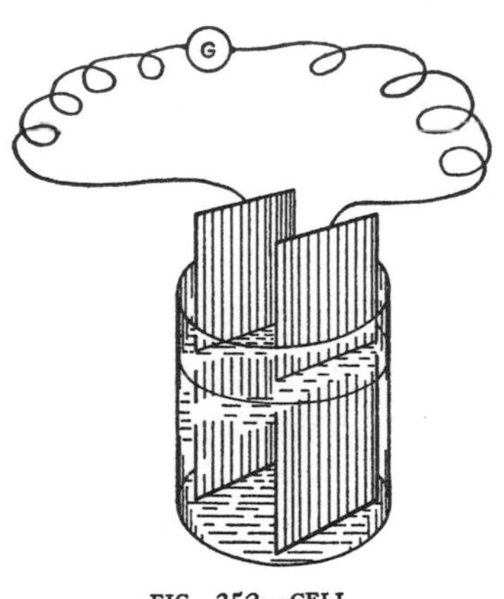

FIG. 352—CELL

Polarization.—After the current has run for some time in the cell as previously described the strength will become very much weaker, but if the copper plate be removed and wiped, then reinserted, the current will be as strong as ever. From this it is learned that the hydrogen bubbles collect on the copper and form an insulator, so that the chemical action is retarded. This forming of hydrogen bubbles is known as **polarization,** and in a good cell there must be some means to check it.

The various forms of cells now in use differ from the above only by using different electrodes and having some method for checking polarization.*

Dry cells.—Dry cells differ from liquid cells only in that the exciting fluid is formed into a jelly or held in suspension by some absorbent such as sawdust or pith.

In the common commercial type the zinc element is in the form of a cylinder and holds the exciting fluid and carbon. The ends of the cylinder are generally sealed with wax. The following proportions by weight will make a very good cell: 1 part zinc oxide; 1 part sal ammoniac; 3 parts plaster; 1 part zinc chloride; 2 parts water.

Heating effect of an electric current.—Owing to the resistance to an electric current passing through a conductor, heat is developed. If the current is small and the cross-section of the conductor large the amount of heat developed will hardly be noticeable, but if the current is strong and the conductor small in cross-section, the latter will soon become hot, often red hot, and sometimes melt down. It is due to this heating effect that many machines are burned out, and it is also due to this same effect that more machines are saved.

Fuse.—If a piece of copper wire is connected in series with one of lead and a current sent through them the lead will melt down at a little over 600° F., but it

*For discussion of commercial cells see any text book on physics or elementary electricity

will require a temperature of nearly 2,000° to melt the copper.

Because lead melts at such a low temperature it is used as a fuse. A fuse consists of a leaden wire connected in series with the circuit it is to protect, and when the current becomes too excessive the lead melts out and thus opens the circuit. Fuse wires, as they are called, are always labeled with the number of amperes they are supposed to carry.

Magnetic properties of coils.—Let a wire carrying a current be formed into a small single coil and bring a compass close to it. When the compass is on one side of the coil it will be noticed that the N pole is attracted and the S pole repelled. Change the compass to the other side and the reverse will be found true. Now reverse the direction of the current and it will be found that the needle acts in just the opposite manner. From this it is learned that the electric coil is simply a flat disk magnet with a N and a S pole, the same as any other magnet.

Electromagnet.—When instead of forming the wire which carries the current into a single loop the wire is formed into several loops in the shape of a helix, a compass brought into its field will produce the same actions of the needle as in the single loop, only they will be much more violent. Now, if a soft iron bar, commonly known as a core, be placed within the helix, a very strong magnet known as an electromagnet will be formed. The lines of force of such a magnet are identical with those of the bar magnet. Hence, if the electromagnet is constructed so that the lines of force can remain in iron throughout their entire length, the magnet will be much stronger. For this reason electromagnets are made in the horseshoe form as shown by Fig. 353.

Electric bell.—The electric bell is a simple application of the electromagnet. The current enters at *A* (Fig. 354), passes through the horseshoe magnet *B*, over the closed circuit breaker *C*, and out at *D*. The instant the circuit is completed through the coils a magnet is formed, which attracts the armature *E*, and rings the gong *F*. But as soon as the armature is drawn down

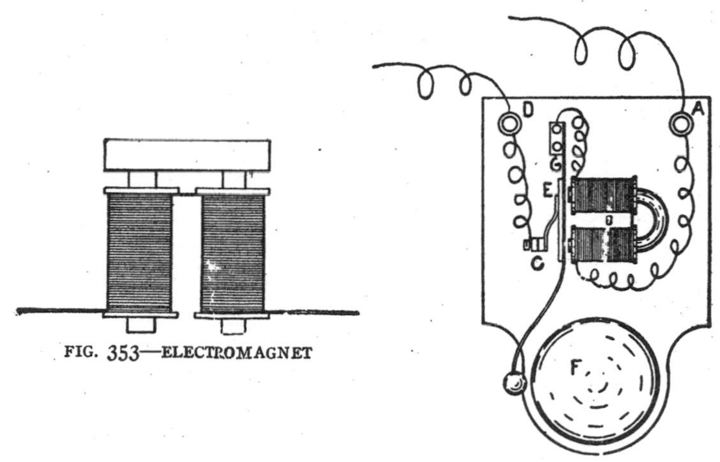

FIG. 353—ELECTROMAGNET

FIG. 354—ELECTRIC BELL

against the poles of the magnet the circuit is broken at *C*, hence the current stops flowing and the magnet becomes nil. As soon as the magnet has no strength the force of the spring *G* draws the armature back and makes contact at *C* again, and the operation is repeated.

Electromagnetic induction.—In a previous paragraph it has been shown that there is a magnetic field surrounding all electric currents. If a wire be arranged so as to form a closed circuit and then moved across a magnetic field a reverse action of that explained above will take place. In other words, if a closed circuit be moved through a magnetic field a current will be set up. This is the most important part of electricity, for upon it is based the operation of nearly all forms of commercial electrical machinery.

Currents induced in a coil by a magnet.—A sensitive galvanometer is connected in a circuit with a wire (Fig. 355) in such a manner that the galvanometer is not affected by the magnet and yet the wire can come into the magnetic field. If that part of the wire between *A* and *B* be very quickly moved down across the field the galvanometer needle will be deflected. When the needle comes to zero and the wire is moved across the field in the opposite direction the needle is again deflected, but the opposite way. If the wire be moved into the magnetic field and held still the needle will come to zero and remain there until the wire is set in motion. Again, if the wire is moved back and forth across the magnetic field the needle will vibrate back and forth across zero, showing that there is a current but an alternating one. When the backward and forward motions of the wire have become fast enough the needle of the galvanometer will practically stand at zero, only giving enough vibration to show that there is an alternating current affecting it. By trial the following results will be obtained:

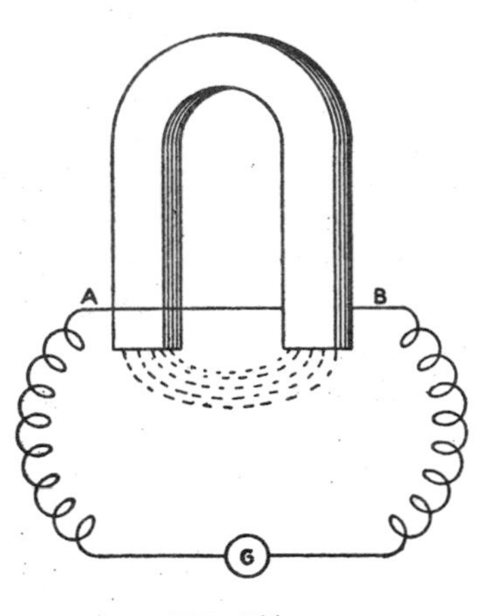

FIG. 355

1. When the magnet is moved and the wire held stationary the same results are noted.

2. When the position of

the poles of the magnet is reversed the current is also reversed.

3. When an electromagnet is used in place of the permanent one the same results are noticed.

4. The induced current is produced by the expenditure of muscular energy and does not weaken the magnet.

5. When the wire is moved so as to cut the magnetic lines of force at right angles the momentarily induced current is greatest.

6. The direction of the lines of force is at right angles to the direction of the current in the wire.

Factors upon which the value of induced E.M.F. depends.—If the wire in Fig. 355 be very quickly moved across the magnetic lines of force the galvanometer needle will deflect farther than when the wire is moved slowly. Also, if two magnets with their similar poles together are used instead of one and the wire is moved at the same velocity as previously the needle will have a greater deflection. Again, if a coil of wire be used instead of a single one the deflection of the needle will be greater. Hence it is obvious that the induced E.M.F. is dependent upon and proportional to the number of magnetic lines cut, the speed or rate at which they are cut and the number of wires cutting them.

Currents induced in rotating coils.—Instead of cutting the magnetic lines of force of a strong magnet with a single wire let them be cut with a coil of 400 or 500 turns. Let the coil be small enough so it will rotate between the poles of a horseshoe magnet. With the coil at right angles to the plane of the poles rotate it 180° and note the direction of deflection of the galvanometer needle. Rotate the coil the other 180° and bring it to the position from which it started and again note the direction of the deflection of the galvanometer needle.

The needle shows that a current has been induced which has two directions of flow during each revolution of the coil. This induced current is produced in exactly the same manner in which currents are produced by dynamos.

Dynamos are machines for converting mechanical into electrical energy. They cannot develop energy but simply change the form of the energy delivered to them. Since they cannot develop energy, the amount of current delivered by them is wholly dependent upon the amount of mechanical energy supplied. In principle the dynamo consists of two parts: a magnetic field made up of electromagnets and a number of coils of wire wound upon an iron core forming an armature.

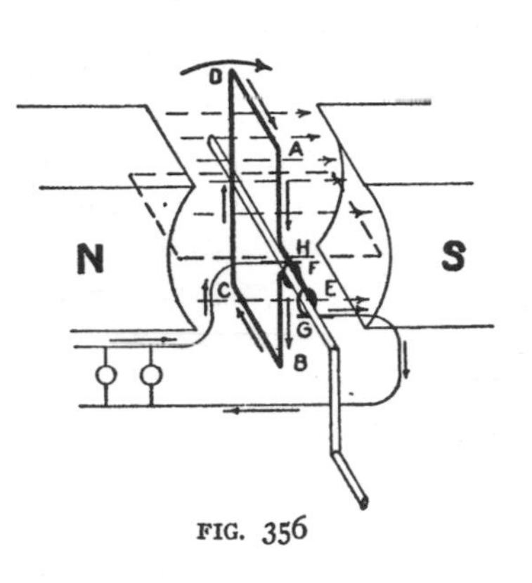

FIG. 356

Simple alternating-current dynamo.—Consider the single loop of wire *ABCD* (Fig. 356) as the armature and the poles N and S as the magnets of a dynamo. With the armature in the position it is shown there is no current developed. The armature is for the instant moving parallel to the magnetic lines of force and consequently is cutting none of them. As the armature moves from a position perpendicular to the lines of force to a position parallel to them, the number of lines it cuts increases until it reaches the perpendicular position, and from then on until it has traversed 180° the number of lines cut decreases until none are cut. From this it is obvious that with the armature in the first and last positions no current is produced and when the armature is cutting the greatest number of lines of force the current is at a maximum. When the armature is turned through the remaining 180° of the revolution the same action takes place. As the side *AD* moves down the current flows in the direction indicated, but as the side *BC* moves down it is reversed. Hence for one half of the revolution the current flows in one direction and for the other half it flows in the opposite direction. One end of the coil is attached to the metal ring *E*, and the other end is attached to the ring *F*. Both rings are fixed to the shaft, so they rotate with it.

Brushes C and *H* are in continual contact with the rings, so the current is taken from them and carried over the circuit.

Armature.—It might be assumed that the iron part of an armature of a dynamo is only to carry the numerous wires which are used for cutting the magnetic lines of force, but this is not the only use for the iron core. The iron carries the magnetic lines of force very much better than they travel through air, and for this reason the

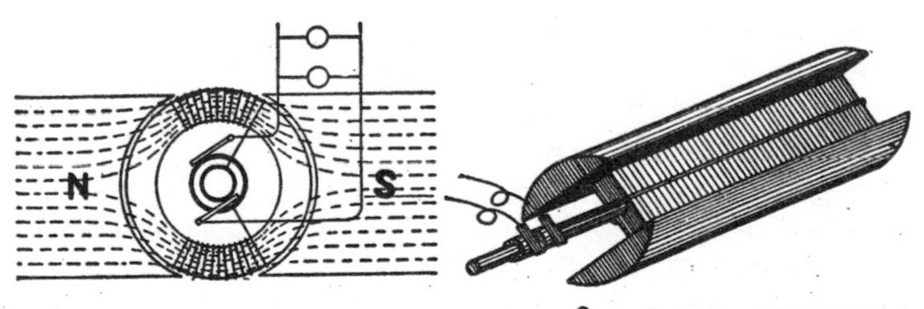

FIG. 357 FIG. 358—MAGNETO ARMATURE

air space between the fields is as nearly filled with the armature as possible. Fig. 357 shows the path of the magnetic lines through a ring armature.

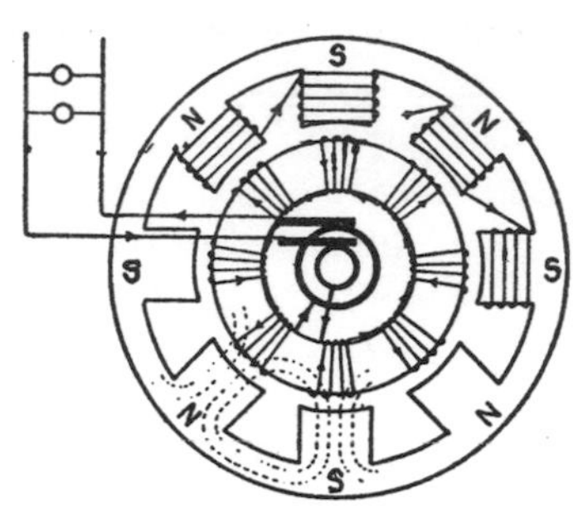

FIG. 359—SYSTEM OF WIRING FOR A MULTIPOLAR ALTERNATOR

638. Magneto alternator.—Fig. 358 shows a magneto armature with the wires off. This is probably the most simple commercial electrical-current generator used. It is only applicable for such uses as cigar lighters, telephone calls and line testers. For large purposes it is too inefficient.

Multipolar alternator.—The number of alternations in a dynamo as just described is 4,000 a minute with a speed of 2,000 revolutions a minute. This speed is as high as advisable, but the number of alternations is only about half as high as is considered good practice. For this reason large commercial dynamos are built with several poles, as shown by Fig. 359, and the number of revolutions reduced. The dotted lines in Fig. 359 represent the directions and paths of the lines of force. The full lines indicate the windings, and the arrowheads the direction of current. By carefully following out the direction of the induced current it will be seen that the coils passing beneath the north poles have a current set up in them which is opposite in direction to that set up in the coils passing under the south poles. By inspecting the windings it will be noted that the direction is reversed between each set of poles, hence the current set up through the system is the sum of all the currents set up at each pole. As the coils of the armature pass across the points midway between the poles, the direction of current is alternated. The number of alternations to the minute is found by multiplying the number of poles by the number of revolutions to the minute.

Direct-current dynamo.—For a great many purposes it is desirable to have a direct current, that is, one which always flows in one direction the same as a current from a cell. To do this some device must be applied to the dynamo just at the point where the current leaves the armature and before it reaches the external circuit. This device as used in a direct-current dynamo is known as a commutator.

Commutators are practically split rings secured to, but insulated from, the shaft of the armature. They take the place of the accumulating rings of the alternator. Each part of the commutator is insulated from the other parts.

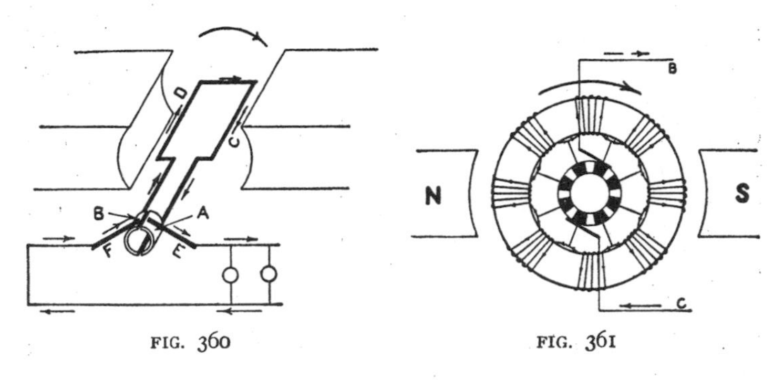

FIG. 360 FIG. 361

Principle of the commutator.—Fig. 360 shows a simple commutator connected to a coil which represents an armature. *A* and *B* are the segments of the metal ring, each of which is connected to the armature. As the armature rotates in the direction indicated by the arrow the current passes off through the side *C*, out over the external circuit through the segment *A*, and in through the segment *B* and side *D*. When the side *D* has passed into the position of side *C*, the current goes out over the circuit in a similar manner. The brushes *E* and *F* must be set so they close contact with each side respectively and make contact with the other side at the instant the current in the armature changes direction.

Ring armature, direct-current dynamo.—A ring armature may be made for a direct-current dynamo by winding on the iron ring a series of coils, the ending of each coil being connected to the beginning of the next. The junction of the two is connected to a section of the commutator. As the number of groups of coils is increased the number of sections of the commutator must also be increased. An eight-coil ring armature is shown in Fig. 361; the direction of current is indicated by the arrows. The induced current from both halves of the armature flows up toward the positive brush *B*, out over the external circuit, back in through the negative brush *C* and through each half of the armature to *B* again. As each coil passes from the field of the N pole and enters the field of the S pole, commutation takes place and the direction of current is reversed. The brushes are located at this point and the current from both sides is conducted off on the same wire. When the brushes pass from one of the commutator bars to another there is an instant when the armature sections are short-circuited; but this is at the instant when these coils are moving parallel to the lines of force, hence there is no current passing through them.

Drum armature, direct-current dynamo.—Instead of winding the armature coils upon an iron ring sometimes they are wound upon a drum. Fig. 362 shows the principle of the drum-wound armature suitable for a bipolar field. Like the windings of the ring armature the coils are in series and both halves are parallel with the external circuit.

Comparison of the drum and ring armature.—By

reference to Fig. 357 of a ring armature it will be noticed that the inside parts of each coil on the armature do not cut lines of force, hence these lines conduct only the current and may be known as so much dead wire. In the drum-wound armature both sides of the coil cut lines

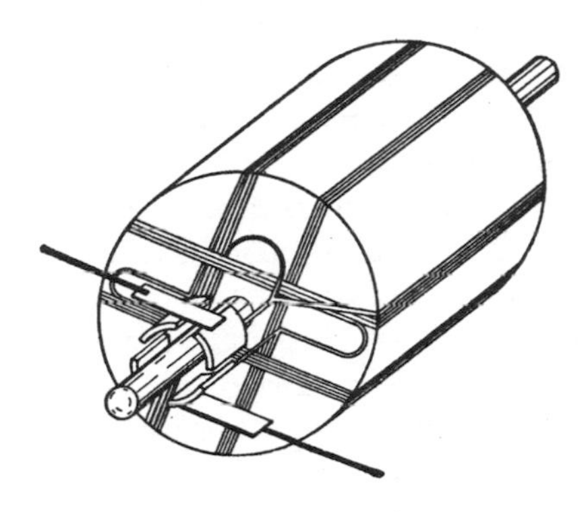

FIG. 362—DRUM ARMATURE

of force and the only dead wire is across each end. Although the drum-wound armature has less dead wire than the ring-wound, it is not as convenient to repair. For this reason high-voltage direct-current arc-lighting dynamos are generally constructed with ring armatures. A combination of the two, which is known as a drum-wound ring armature, is extensively used in practice.

Self-exciting principle of dynamos.—In the earlier types of dynamos the field magnets were always separately excited by either a battery or a magneto. Later it was learned that the soft iron of the field magnet after once being excited retains some of the magnetism. Since then all direct-current dynamos are built on this principle. There is sufficient magnetism remaining in the fields so that when the armature is up to speed it cuts enough lines of force to induce a small current into the circuit around the field coils. This current more highly excites the field magnets until the dynamo soon picks up or establishes its rated E.M.F.

Shunt dynamo.—In the so-called shunt-wound dynamo a small portion of the current is led off from the brushes *bb* (Fig. 363), and through a great many turns of very fine wire which encircle the core of the magnet. In such a dynamo, as the load increases the E.M.F. slightly decreases, and as the load decreases the

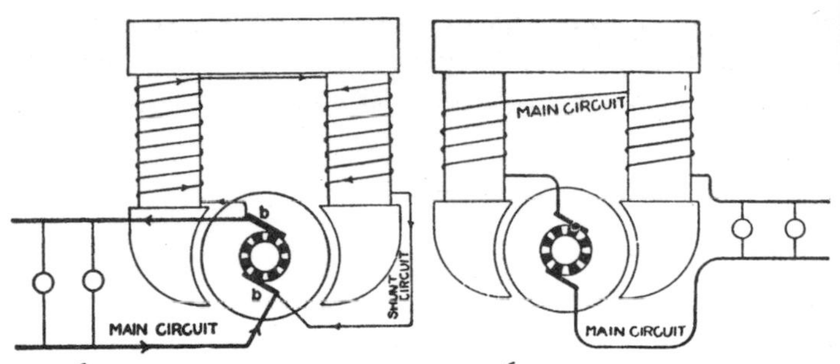

FIG. 363—SHUNT-WOUND DYNAMO FIG. 364—SERIES-WOUND DYNAMO

E.M.F. increases. Hence, if the current fluctuations are great and quite frequent it would keep an attendant occupied to keep the field resistance regulated for the load. (See Fig. 368.)

Series-wound dynamo.—In the so-called series-wound machines the whole of the current is carried through a few turns of very coarse wire which encircles the field magnets (Fig. 364). Since every change of current alters the field magnetizing current, consequently in the current induced in the armature the E.M.F. at the brushes will vary with every change of resistance in the external circuit.

Compound-wound dynamo.—In the compound-wound machines there is both a series and a shunt coil surrounding the cores of the field magnets. This style of machine is designed to give automatically a better regulation of voltage on constant-potential circuits than is possible on the shunt-wound machines, and yet possesses the characteristics of both the series and shunt machines. Like the shunt machine a part of the current is shunted from the brushes and around the magnet cores, also the external circuit is thrown around these cores. These machines are designed especially for conditions in which the load is very variable, as street car work, incandescent lighting and for commercial power purposes.

Classification of dynamos. — Dynamos may be classified according to their mechanical arrangement as follows:

1. Stationary field magnet with revolving armature.
2. Stationary armature with revolving field magnet.
3. Stationary armature and stationary field magnet with revolving core.

They may also be classified by mechanical designs as follows:

1. Direct-current machines.
2. Alternating-current machines.

And by electrical arrangement as

3. Shunt-wound.
4. Series-wound.
5. Compound-wound.

Armatures.—The armature core introduced into the magnetic circuit to help lower the reluctance is also an electrical conductor, and when rotated in a magnetic field will have currents set up within itself. These currents are independent of the external circuit, hence are

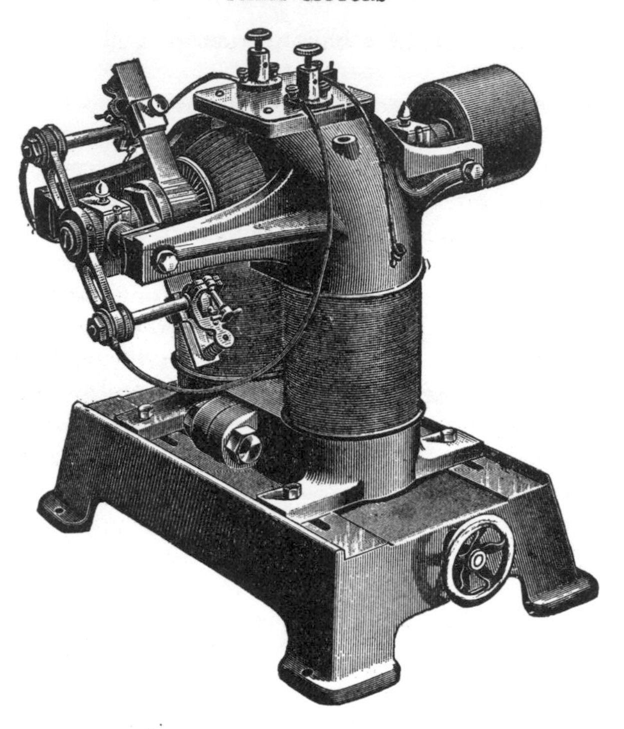

FIG. 365—BIPOLAR DIRECT-CURRENT DYNAMO

a loss. They are known as eddy currents and the loss is termed **eddy current** loss. Fig. 366 shows a section of a solid armature and the direction of these currents. Not only do these currents create a loss themselves but they heat the armature windings and thus increase the armature resistance. If these large eddy currents can be broken up into smaller ones the loss will not be so great. To break up these eddies armatures are now generally built up of a large number of sheets of iron with insulation between the sheets. The insulation used for this purpose is generally a coat of rust or a sheet of tissue paper.

Hysteresis.—Another source of loss in an armature is due to the fact that every time the current alternates the polarity of the magnetism is reversed. If the armature is making 2,000 revolutions a minute and there are two alterations in each revolution there would be 4,000 alterations of the magnetism. This causes heat in the armature which is not accounted for in the external circuit, hence is a loss. Not only is there loss by heat in the armature, but the heat acts on the coils and increases the resistance in them and creates another loss. The loss in an armature due to these alterations of magnetism and the heat produced thereby is known as hysteresis loss.

Insulation of an armature.—The insulation of an armature is probably the most essential part of a dynamo. After it is put on in the various places where it is needed it must be baked and all moisture evaporated out of it. After an armature is thoroughly prepared for use it is generally tested for poor insulation. The potential difference for the test is about eight times as much as the armature is expected to carry. If there is any place where the electricity breaks through the insulation it is detected by means of a sensitive galvanometer.

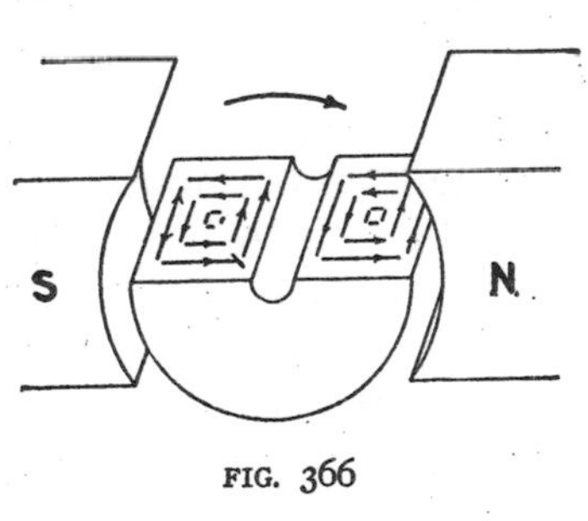

FIG. 366

Capacity of dynamos.—It would seem that the amount of current that a dynamo could produce might be indefinite if enough power be supplied. This is true in a certain sense, but there is a limit and this will appear in one of the following ways:

By poor regulation of voltage.—An overload will cause an excessive drop of the E.M.F. at the machine. This will decrease the potential difference at the brushes and cause a weak current over the line.

By excessive heating.—The heat from an armature increases four times for each doubling of the current. At this rate the armature would soon become red hot. It would work at a little less than red heat, but even this much heat would break down the insulation. The armature should not become warmer than 212° F., and the general custom is not to run it at a higher temperature than 70° above the surrounding air.

Commercial rating of dynamos.—Dynamos are rated according to the number of kilowatts they will carry in the external circuit without excessive heating. For example, a person calls for a 60 K. W. 110-volt generator. This means that he desires a machine which will deliver 60 K. W. to the external circuit and maintain a potential difference of 110 volts across the brushes. Owing to losses in the machine such a machine may develop 63 K. W. and still have only 60 K. W. available for use in the external circuit.

Efficiency of dynamos.—The efficiency of a dynamo is the ratio of its electrical output to the mechanical energy exerted upon it. For a 1 K. W. machine it is only about 50 per cent, and in generation of several thousand kilowatts it is about 95 per cent.

Sparking at the commutator.—Sparking at the commutator is the most serious trouble the attendant will have with a dynamo, provided he keeps all other parts clean, and the insulation does not break down or the machine become short-circuited. There are several causes for a dynamo to spark, some of which are:

1. Brushes not set at neutral point. This can be remedied by

working the brushes back and forth until the proper position is located.

2. Brushes not spaced according to commutator bars. The commutator bars should be carefully counted and the brushes accurately set between them.
3. Brushes do not bear against commutator with sufficient pressure.
4. Brushes do not bear on the commutator with a perfect surface.
5. Collection of dirt and grease which prevents good contact of the brushes on the commutator.
6. A high or low commutator bar which causes poor contact.
7. Commutator not worn perfectly round, consequently poor contact with the brushes.

Repairing a dynamo.—If the insulation breaks down, a wire burns out or the commutator becomes worn out of round, an expert should be called in, and generally the defective part will have to be sent to the factory for repairs. Sometimes a good machinist can put the armature in a metal lathe and turn it down round. A good man with a file can work down a high bar, and holding a piece of sandpaper on the commutator while it is in motion will clean it of all oil and dirt.

MOTORS.

Comparison with a dynamo.—A dynamo is a machine for converting mechanical energy into electrical. An electrical motor is just the reverse; it is a machine for converting electrical energy into mechanical. Any machine that can be used as a dynamo can when supplied with electrical power be used as a motor. Dynamos and motors are convertible machines; thus the various discussions will apply as well to the motor as to the dynamo.

Principles of the motor.—It has been shown that when a coil of wire is placed in a magnetic field and rotated an electrical current is produced. If the opposite of this is done, i.e., if a current is passed through the coil, the coil will tend to rotate. This is the principle of the electric motor: instead of taking a current off of the armature, one is put into it and at the same time sent through the fields. The current passing through the

FIG. 367—MULTIPOLAR MOTOR

fields induces magnetism in them; the lines of force produced by this magnetism draw on the armature and cause it to revolve. By studying Fig. 356 it will be noticed that the coil will revolve until the plane of the coil is parallel to the lines of force, and then stop. This same condition would take place in the motor if it were not for the commutator. Just at the instant the coil is brought to the position to stop, the commutator changes the direction of the current and the turning effect is thrown to the other side and the armature moves on.

Counter electromotive force of a motor.—The armature wires of a motor rotating in its own magnetic field cut the lines of force as if the motor were being driven as a dynamo, consequently there is an induced E.M.F. in them. The direction of this induced E.M.F. is opposite to that of the applied pressure. Such an induced E.M.F. is known as counter electromotive force and is an important property of the motor. A motor without load will run with sufficient speed that its counter electromotive force will very nearly equal the applied pressure. The counter E.M.F. will never be as great as the applied force. There will always be a difference between these, equal to the loss due to resistance in the motor armature. The power of a motor increases as the counter E.M.F. decreases until the counter E.M.F. is one-half of the applied E.M.F., then the power of the motor decreases. The maximum power of a motor is reached when the counter E.M.F. is one-half of the applied E.M.F.

Losses of a motor.—The losses of a motor, like those of a dynamo, are due to resistance in the armature friction, eddy currents and hysteresis.

Operating motors.—The resistance in the armature of a motor is so low that if a motor were directly connected to the supply mains, too great a current would flow through it before a counter E.M.F. could be set up, consequently the machine would be practically short-circuited and the windings damaged. For this reason a rheostat known as a starting rheostat is inserted into the armature circuit of a shunt motor. To start the motor, switch *A* (Fig. 368) is closed, and this throws the current into the fields and excites them; then the arm is moved over the starting box to point one, and when

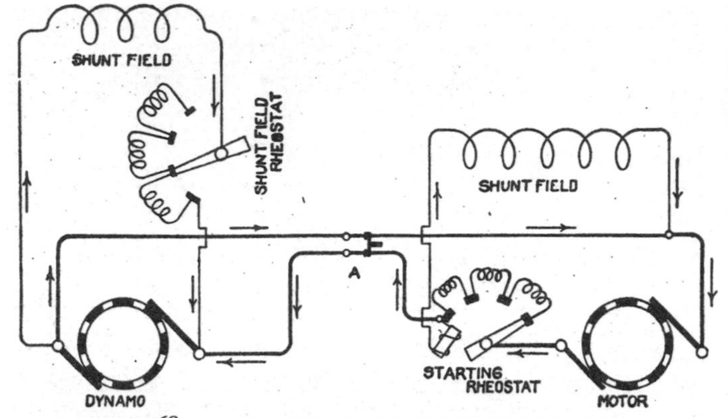

FIG. 368—WIRING SYSTEM FOR DYNAMO AND MOTOR

the motor has attained its speed for this point it is moved on up to point two, then three, and so on until the last point is reached and the motor is directly connected to the feed wire. To stop the motor, switch *A* should be opened, and if the arm *B* is not an automatic shifter, it should be thrown back to its original position ready for starting the next time. Most of these arms are now made so they work against a spring, and when the last point is reached an electromagnet attracts the arm sufficiently to hold it in position; then when the circuit is broken the magnet loses its attraction for the arm, and the spring draws it back.

The electric arc.—When a current of from 6 to 10 amperes under a pressure of about 45 volts is passed through two rods of carbon with their ends first in contact, then gradually drawn apart to a distance of about 1/8 inch, a brilliant arc of flame is established between them. This arc, known as the electric arc, is made of a vapor of carbon. As the current passes across the contact points the high resistance produces enough heat to

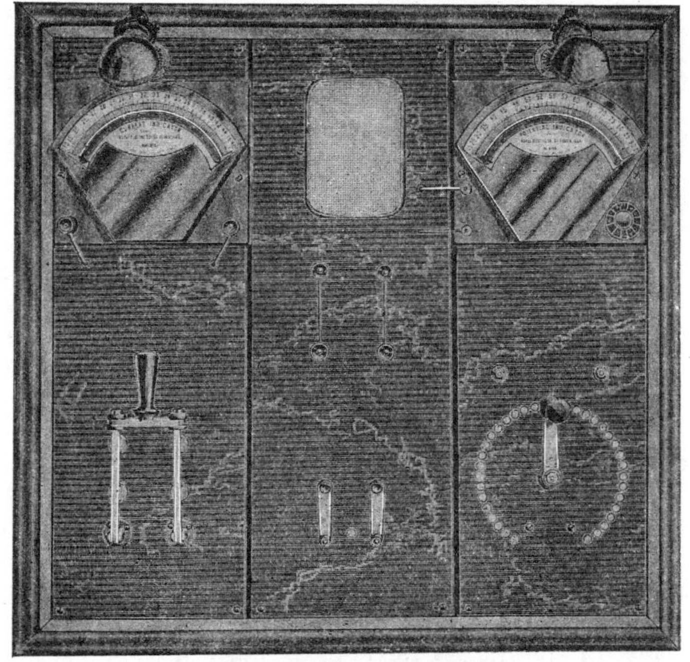

FIG. 369—COMMERCIAL SWITCHBOARD

disintegrate the carbon and cause it practically to boil; this boiling throws off a vapor which is a conductor of electricity and as a consequence conducts the current across the gap. The temperature of the arc at its hottest point is about 3,500° C., which is about twice the temperature required to melt platinum, the most refractory of metals.

Arc lamps are rated according to the watts consumed. They generally range from $6 \times 45 = 270$ watts to $10 \times 45 = 450$ watts. About 12 per cent of the energy supplied to an arc light really appears as light; the rest goes to produce the heat evolved.

Since the carbons of the arc lights are constantly wasting away there must be some device to regulate the distance they are from each other and to work automatically to keep them at this distance. An ingenious appliance of electromagnets and clutches accomplishes this action and is explained in any book upon electric lighting.

Incandescent lamps.—It is on the principle of the heated wire that we get light from the incandescent lamp. Referring to Fig. 370, connections are made with the lamp at *A* and *B*. At *CC* are bits of platinum wires attached to the carbonized filament *D*. *E* is the highly exhausted globe. If the carbonized filament were in the air, the intense heat created within it due to the resistance of the current would immediately burn it up, but since it is in almost perfect vacuum, it will last from 600 to 800 hours. Even at the end of this period the filament does not always break, but it becomes so disintegrated that the candle power is low and further use is not satisfactory.

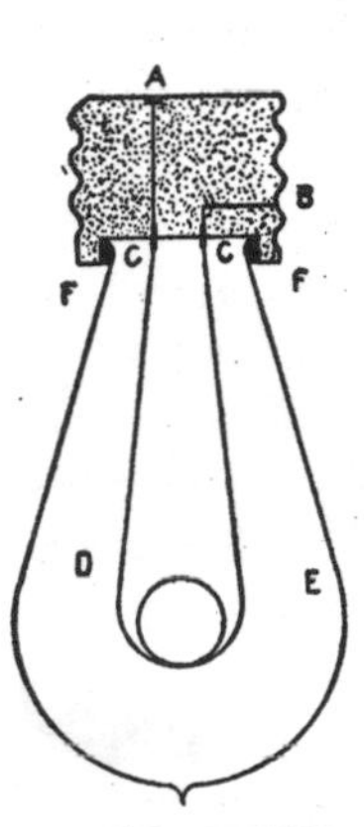

FIG. 370—INCANDESCENT LAMP

Commercial rating of incandescent lamps.—Before a lamp is put upon the market it is compared with a lamp of known brilliancy. While it is being compared with the standard lamp, measurements of its voltage and current are made. After this is done the lamp is labeled with the voltage it carries, its candle power and watts consumption. A 16 C.P. 60-watt 110-volt lamp will require

$$C = \frac{W}{E} = \frac{160}{110} = .55+ \text{ amperes.}$$

Lamps are usually made for circuits of 50 to 60 volts, 110 to 115 volts and 220 volts with constant potential.

A 16 C.P. lamp requiring 55 watts on a 50-volt circuit will take about one ampere; on a 110-volt circuit it will take 0.5 ampere; on a 220-volt circuit about 0.25 ampere. A lamp should not be subjected to a voltage higher than its rating; the filament is not made for it and will soon give out.

The efficiency of a lamp is proportional to the ratio of the number of candles it will produce to the number of watts it absorbs. A high efficiency is 3 watts per candle power, and the average efficiency is 3.5 watts candle power. High-efficiency lamps are used where the pressure is very closely regulated or cost of power is high, and low-efficiency lamps are used where there is not such close regulation and power is less expensive.

Potential distribution in lamp circuits.—Incandescent lamps are usually operated from low-voltage constant-potential circuits. Where lamps are supplied with current from a street car circuit, which generally has a potential of 500 volts, they are grouped in multiple series; i.e., 5 100-volt lamps or 10 50-volt lamps will be connected across the mains. In a series circuit the drop on the lead wires does not interfere with the regulation of the voltage at the terminals, but in a parallel circuit this drop is an important factor and requires that the lamps be distributed and the size of wire proportioned so that each lamp receives about the same voltage. For example, consider 100 220-volt lamps to be connected at distances along a pair of mains which extend 500 feet from a generator which has a potential difference of 225 volts at the brushes. The lamps nearest the dynamo will receive a greater potential than their rated capacity and will often burn out, while those farthest from the dynamo will not receive potential equal to their capacity, hence will burn dimly. In order to overcome this, centers of distribution are laid out in wiring construction and groups of lamps are fed from these centers Fig. 371. Feed wires are run from the generators to these centers and a constant potential is kept in them by regulation at the generator. Sets of mains are run from these centers, and then submains are led off from these mains to supply subcenters of distribution. To these subcenters lead wires to the lamps are connected. In this system of wiring it does not matter if there is a fall of potential of 20 per cent, between lamps and generators, for the fall is alike in all. For example, a voltmeter across the brushes of a generator shows 225 volts, one at the main center of distribution shows only 218 volts, one at the subcenters shows only 212 volts and one across the terminals of the lamp shows only 210 volts. But since there has been the same number of divisions and subdivisions the P.D. of all of the lamps is the same.

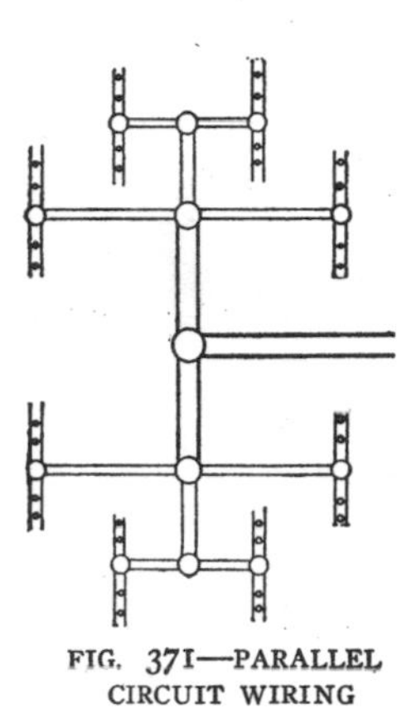

FIG. 371—PARALLEL CIRCUIT WIRING

Calculations for incandescent wiring.—To find the size of wire for carrying a certain current, let

C. M. = circular mil area of wire,

K = 10.79 = resistance 1 mil foot of copper wire.

L = length of circuit in feet,

C = current in amperes,

E = volts drop on the line.

In the formula,

$$C.M. = \frac{K \times L \times C}{E} = \frac{10.79 \times L \times C}{E}.$$

After obtaining the circular mil area, this must be compared with a wire table to get the number of wire to use.

Example.—Fifty 55-watt 110-volt lamps are connected in parallel to a center of distribution located 100 feet from a dynamo which generates 112 P.D. By measurement the potential at the point of distribution is 110 volts. What size wire is required for the feeder?

To find amperes to be conducted.

$$C = \frac{W}{E} = \frac{55}{110} = 0.5 \text{ per lamp.}$$

$$0.5 \times 50 = 25 \text{ amperes for all lamps.}$$

$$112 - 110 = 2 \text{ volts drop on line.}$$

$$C.M. = \frac{K \times L \times C}{E} = \frac{10.79 \times (100 \times 2) \times 25}{2} = 26{,}975.$$

C. M. = circular mil area.

K = 10.79.

L = 100 × 2 = 200 feet.

C = 25.

By comparison with the wire table (670) the next larger size than 26,975 is B. & S. No. 5.

Wiring calculations for a motor.—To find the size of wire to transmit any given horse power any distance when the voltage and efficiency are known.

$$C.M. = \frac{H.P. \times 746 \times L \times 10.79}{E \times e \times \%M}.$$

E = voltage required by motor.

e = drop on line.

H. P. = horse power of motor.

% M = efficiency of motor in decimals.

Example.—What size of wire is required to conduct current to a 220-volt 6 H.P. motor located 175 feet from the dynamo? The drop on the line is to be 6 volts and the efficiency of the motor 80 per cent:

$$C.M. = \frac{H.P. \times 746 \times L \times 10.79}{E \times e \times \%M}.$$

$$= \frac{6 \times 746 \times 175 \times 2 \times 10.79}{220 \times 6 \times .80}.$$

$$= 15{,}984 \text{ C. M.}$$

$$= \text{No. 8 B. \& S.}$$

To find the current required by a motor when the horse power, efficiency and voltage are known.

$$C = \frac{H.P. \times 746}{E \times \%M}.$$

Example.—What current is furnished to the motor in the previous problem?

$$C = \frac{H.P. \times 746}{E \times \%M},$$

$$= \frac{6 \times 746}{220 \times .80},$$

$$= 25.4 \text{ amperes.}$$

INDUCTION COILS AND TRANSFORMERS

Self-induction.—Self-induction is defined as the cutting of a wire by the lines of force flowing through the wire. When a current begins to flow through a wire magnetic whirls spring outward from the wire and cut it. This cutting of the wire with only its own magnetic lines of force induces an E.M.F. for an instant. But the E.M.F. which it does induce has an opposite direction to the E.M.F. which causes the current to flow. Hence the E.M.F. will be retarded for an instant by its own induced E.M.F. and will not flow until this is overcome. When the current flowing through the wire is stopped the lines of force again cut the wire but in an opposite direction, hence this time they tend to retard the cessation of flow of the current. The effects of self-induction are rarely noticeable in a straight wire, but when the wire is wound into coils in the form of a helix the magnetic field of every turn cuts many adjacent turns and the E.M.F. is greatly increased, being proportional to the current, the number of turns and the magnetic lines through the coil. If an iron core is placed within the coil the effects of self-induction are very much greater. By snapping the wires from a battery after passing through such a coil as described above a brilliant spark will be produced. This is the simple coil (Fig. 372) used in make-and-break ignition on gasoline engines.

667. Induction coil.—If two coils entirely separate from each other be wound around an iron core and connected up as in Fig. 373 every time the current is started in coil *a* there will be a deflection of the galvanometer needle in *b*. If the current is broken in *a* the needle *b* will again be deflected, but in an opposite direction. From this it is seen that the magnetic lines of force which surround the wire in coil *a* induce a current in the coil *b*.

This is the principle of the induction coil, a diagram of the connections being shown in Fig. 374. The circuit leading from the batteries to the inside of the coil is known as the primary and the circuit wound on the out-

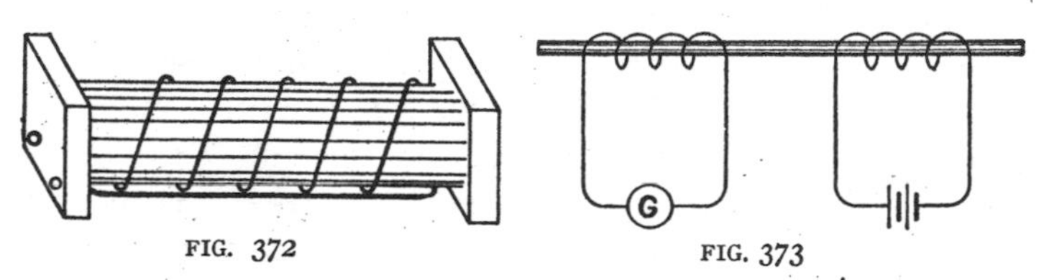

FIG. 372

FIG. 373

side of this is known as the secondary. The primary induces the current in the secondary, and if the secondary circuit has more turns of wire than the primary it will have a correspondingly greater E.M.F., in other words, the difference in E.M.F. of the two circuits varies directly with the difference in the number of turns in the wire of the two. Since the induced E.M.F. is set up only as the current is made or broken, an automatic device *A* is connected into the primary, whose action is identical with the circuit breaker of an electric bell. In induction coils this, however, is generally known as a buzzer.

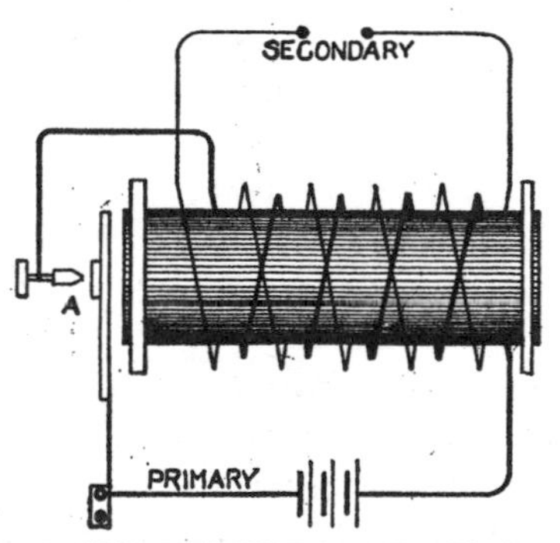

FIG. 374—PRINCIPLE OF THE INDUCTION COIL

The induction coil is used with jump-spark ignition, on gasoline engines. For this work the spark requires such a high E.M.F. that the primary consists of only a few turns of coarse wire, while the secondary consists of several thousand turns of fine wire.

Transformers.—Where alternating currents are used for electric lighting, to make the cost of transmission a minimum a voltage of 1,100 to 2,200 or even higher is used; this is far too high to be taken into houses and so a transformer is connected into the circuit. A transformer is identical with the induction coil with the automatic circuit breaker removed. A transformer, however, usually decreases the E.M.F. instead of increasing it. This is done by having the primary enter the coil on a large number of turns and the secondary pass off on a few turns. Since the current is alternating in action, it takes the place of a circuit breaker.

Copper wire table.

Gauge, A. W. G. B. & S.	Diameter, Inches	Area, Circular Mils	Weight Pounds per 1,000 Feet	Length, Feet per Pound	Ohms Resistance per 1000 Feet
0	0.3249	105,500	319.5	3.130	0.09960
1	0.2893	83,690	253.3	3.947	0.1256
2	0.2576	66,370	200.9	4.977	0.1584
3	0.2294	52,630	159.3	6.276	0.1997
4	0.2043	41,740	126.4	7.914	0.2518
5	0.1819	33,100	100.2	9.980	0.3176
6	0.1620	26,250	79.46	12.58	0.4004
7	0.1443	20,820	63.02	15.87	0 5048
8	0.1285	16,510	49.98	20.01	0.6367
9	0.1144	13,090	39.63	25.23	0.8028
10	0.1019	10,380	31.43	31.82	1.012
11	0.09074	8,234	24.93	40.12	1.276
12	0.08081	6,530	19.77	50.59	1.610
13	0.07196	5,178	15.68	63.79	2.029
14	0.06408	4,107	12.43	80.44	2.559
15	0.05707	3,257	9.858	101.4	3.227
16	0.05082	2,583	7.818	127.9	4.070

CHAPTER VIII

THE FARM SHOP

Necessity.—There is no farm so small but a farm shop would be of value. For small farms there should not be many tools, but there is seldom a year when a small investment in a bench with a vise and a few tools would not return to the user a good dividend. It is not alone the amount of money which can be saved by doing a large per cent of one's own repairing, but it is the time saved in emergencies.

Often breakages occur with farm machinery which, if the tools are at hand, may be repaired in much less time than is required to take the broken parts to a repair shop where the job must wait its turn with others equally urgent. There are times when farm work is very pressing and a delay of a few hours means a loss of many dollars in wasted crops.

Not only is there a loss by not having a shop for urgent repairs, but there are rainy and disagreeable days, when men do not relish working outside, that can very profitably be put in working in the shop.

Use.—The idea is prevalent that only skilled mechanics can do work in a shop. Of course this is true in a great many instances where the work is difficult, but there are more times when the work is such that a man with only ordinary mechanical ability can do it. The farmer should not attempt to point plows, weld mowing machine pitmans and do such work until he has achieved skill. However, he can tighten horseshoes, repair castings, etc., as well as do carpentry work.

Location.—The location of the shop depends greatly upon circumstances and taste. If the shop is equipped with only a work bench and the tools which go with it, it can be built in the barn, or a part of the machine shed be used. In fact a suitable place can be arranged almost anywhere. To locate a shop with a forge in the equipment is a little more trouble. It should be a separate building and far enough away from the other buildings so that in case it should catch fire the other buildings could be saved. Should the owner of a farm shop be fortunate enough to possess a gasoline engine or some similar source of power, the engine can very handily be placed in a room adjoining the shop and a shaft run one way into the shop and another way into the granary where the sheller and grinder may be located.

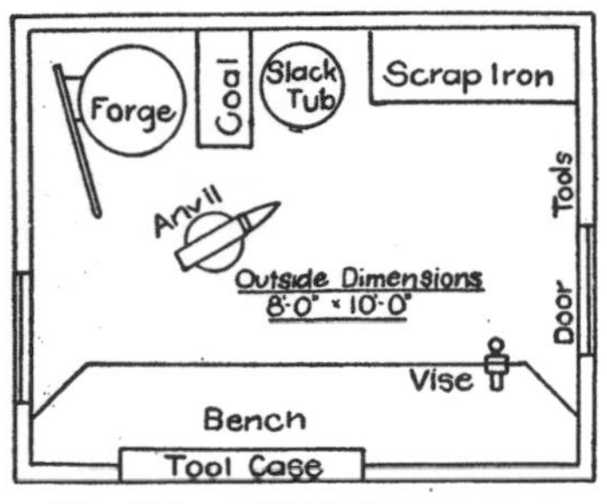

FIG. 375—ARRANGEMENT OF A SMALL SHOP

Construction.—That part of the shop floor about the forge and anvil should be of earth or concrete, and if concrete be used in this part it might as well be extended over all the floor space. The material and design of the outside of the shop should conform to the style of the other buildings about the place.

Size.—The size of the shop should conform to the size of the farm and a man's ability as a mechanic. A small farm does not require as well equipped a shop as a large one. A farm close to town does not require as large a shop as one several miles in the country. A man who is inclined to handle tools more or less will make very much more use of a shop than a man who will

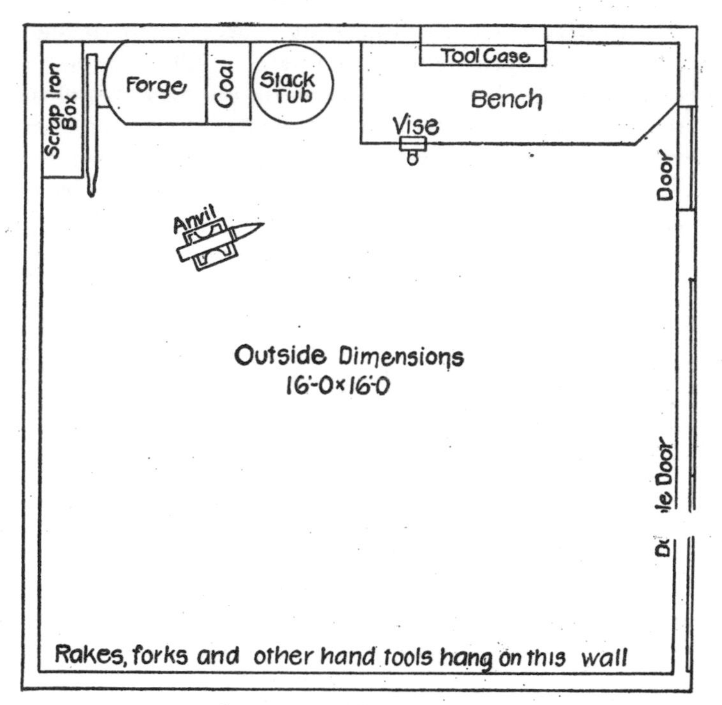

FIG. 376—ARRANGEMENT OF A LARGE SHOP

use it only when dire necessity requires, consequently the man who uses the shop frequently needs a larger one than the man who very seldom enters it. A shop with a floor space of 8 × 10 is large enough for a bench with a few hand tools and a small portable forge.

If one desires to have his shop large enough so that a wagon can be run in for repairs it should be about 16 × 16 feet. It might seem that this would be a waste of space, but that part of the shop where the wagons stand for repairs can be used for a wagon shed all the rest of the time.

Equipment.—The following is a list of tools suggested for a farm of 160 to 320 acres. The cost of the wood tools is from $15 to $20, according to grade, and the cost of the forge tools from $25 to $35. The anvil referred to in this list is cast iron with steel face; if a wrought-iron anvil with a steel face be substituted for it an addition of about 5 cents for each pound weight should be added.

NOTES

NOTES

NOTES

NOTES

NOTES

NOTES

NOTES

NOTES